BRINGING
PRACTICAL
SOLUTIONS
FOR THE
GOLDEN AGE
A CRITICAL EXPOSE OF THE NEW GREEN DEAL

A NEW GREEN DEAL REPORT CARD

2018-2025

VOLUME I

Dr. James R. Boynton

Dedication

This dedication is a bit unusual, but I find this book would not have been possible without the referential materials I got in my research from hundreds of articles written by corporations, companies, engineers, inventors, entrepreneurs, and financiers. Of course, accreditation has been given to the sources and people who wrote them, and I have been advised in emails and phone calls I made to people and associations to learn more about their activities.

One thing that was sorely needed largely comes to the surface: companies and others who rose to the need for solutions to environmental issues rightly pointed out by the NGD people. I had thought our economy had buried the most needed commodity, money, for research and development, but surprisingly, this was not the case. Most of these improvements will actually reduce operating costs.

I fervently believe the unplausible solutions suggested by the NGD folks were untenable at best, such as batteries in planes, fuel issues for vehicles, planes and trucks, and energy requirements. In this book, you will find innovative technology that can actually take the onus off of harmful emissions. Industries worldwide are working to minimize the impacts that NGD folks claim will shorten Earth's habitat.

I am astounded but pleased that these industries have risen to the task of bringing innovative technology to the forefront.

The US atmosphere will benefit from this coming to the forefront and blunting the demands due to the corrective actions being deployed and refined. But I am still disturbed that nations who will not embrace these developments will retard the desired outcomes.

Of course, this will reduce the rampant costs the NGD program is spending, but only if we keep the money in the US. Citizens first, please.

Author note I am going to make myself wildly unpopular by the following statement. The stated final outcomes of the New Green

Deal cannot and will not ever be realized, but you must read through this book to find out. The outcome for the world population isn't rosy."

Important author note:

This book was written and entered the publication process before the President of the United States declared, by executive order, that the New Green Deal program was declared stopped.

The House of Representatives placed this stoppage into the what they call the "Big Bluetiful Bill", passed the bill and sent it to the Senate where it is today 6/18/25 awaiting a confirming vote.

When this vote passes the bill it will go to the President for signature and the New Green Deal will officially be cancelled.

What is in this manuscript relates to all the history and activities of the New Green Deal for you to read and learn about what this program severly cost us in your tax dollars and the damages they did to people and industries that were deemed "not valuable or needing drastic changes according to their philosophy of greening the Earth even if they were ecologically unsafe or dangerous. Not all was in our best interests. Please read this with an open mind.

Dr. James R. Boynton

Acknowledgment

Any acknowledgment would be in vain if it did not give our lord and creator proper credit. He has touched my heart and mind to deliver this manuscript. It is in his name that this manuscript is dedicated to, and our fervent wish is that you are enriched by its content. Thank you.

In 2020, friends encouraged me to pen a manuscript concerning the New Green Deal. They and I did not believe this was the best for our country, and it would literally upset the normalcy of functioning systems at all levels of our society. I authored the book *The New Green Deal: Back to the Stone Age* under the pseudonym Shane Roberts, and we witnessed the fundamental changes of an unmandated, officially unvoted program of destruction, forcing changes upon the public to unfold. The major spending commenced and continued until early 2025 with many critical high-dollar failures (billions, nearing trillions) to their credit.

At the time, I used the symbolism of a Trojan Horse emulating the hidden enemy within with a hidden agenda, not unlike the deceptive Greeks of the Trojan War with Troy, not in the whole light of day.

Unfortunately, from 2020 till 2025, not much has changed. They were not accountable; it seems so to anyone. When have you heard of this group bringing any of this forward with the full approval of the US House or Senate? Fact, even though they are acting like Don Quixote charging at -ha-ha windmills and solar panels, which are nebulous figures of their imagination to solve our power needs and reducing our natural resources to our distraction and killing millions of jobs with the promise of reeducation for those in those occupational fields that really never materialized.

My acquaintances were right, and I thank them profusely for their continued assistance, including this manuscript. I particularly honor Pete, my best friend. I cannot dismiss the moral, professional, literary, and administrative guidance that many writers seem to need, which I got from my wife, Linda. I also want to thank my sister Susan, a highly published author, for keeping me on track with concepts with wizened suggestions; everyone needs a backstop like her.

An especially important author note: This book was written prior to the recent cancellation of the New Green Deal. The facts and informative materials herein support the need for the cancellation of this expensive and largely ineffective program, but one important fact still remains we still do have problems with climate controls we can affect, and a number we cannot, but our job, here in the United States is to take care of our issues. This is a responsibility for all of us. Our future depends on our concentrated stewardship. Thank you.

About the Author

Hon. Dr. James Robert (Jim) Boynton

Jim Boynton holds dual majors in Business and Accounting, with a minor in Computer Science. He holds a Doctor of Philosophy (PhD)in Operations Management. An accomplished international author and speaker, Jim has been widely published in conference journals, periodicals, and editorials and is known for his work globally.

His books, former, present, and pending, are The New Green Deal; Back to the Stone Age (under the pseudonym Shane Roberts) 2020, Genesis of a Nation of We the People, released October 7th, 2024, to the open market and he is, with this new book submission of:

Bringing Practical Solutions for the Golden Age
A Critical Expose of the New Green Deal
A New Green Deal Report Card
2018-2025

Dr. Jim Boynton, we will round out several successful years of publishing key informational manuscripts with more in development for 2025.

Jim has 66 years of varied industry experience in leadership as a college/university professor, consulting, and farming. He has held significant roles, including President and CEO of Focus NH Institute, an educational development organization responsible for developing a four-year innovative college curriculum course approved by the New York Board of Regents. His leadership and teaching positions in international organizations such as APICS and ASQ were instrumental in bringing innovative technology advances to many audiences.

He has contributed to major reforms in New Hampshire educational institutions at the governor's request. A congressman nominated him as a National Educational Standards Board participant. He was instrumental in developing and instructing engineering courses at the graduate level for the University of New Hampshire and developed and instructed business courses at the New Hampshire Community Technical College and McIntosh College.

Jim's political career includes serving five years as a New Hampshire State Representative and County Commissioner of a 360,000-person county. He was also a candidate for the State Senate and actively participated in various community and state initiatives, including water resources and major economic development.

His military service spanned over two decades, including active and reserve duty in the Navy and the Massachusetts National Guard, where he held various senior leadership and operational roles.

Retiring in 2004, Jim moved to Tennessee, where he bred rare Desert Arabian horses until a prior military injury ended his career in 2014. He now resides in a secluded community, enjoying a peaceful life amidst nature and writing.

Table of Contents

Preface

As a former university professor, I couldn't resist issuing a report card on the events of the past few years, specifically from 2018 to 2025. These years have seen some rather intriguing developments, particularly with the Green New Deal's unreasonable expectations and what I felt was out-of-control spending, but industry came to the rescue in innovative ways.

We have witnessed the good and bad interactions in our country after the launch of this economic phenomenon called the New Green Deal, a rebranding of the New Deal from the forties. A newer relaunch in 2024 introduced additional costly changes, especially in the older and new construction housing markets.

Living in the South, I had an interesting conversation about the Green New Deal with a 'dyed-in-the-wool' Southerner. I asked him what he thought about it, and I hope I did his comment justice.

"Nah, it ain't nuthin'. Them people are nuts, ain't gonna mount to nuthin'. Folks don't trust 'em - they're stellin' all our bucks. Did me in? Heck, man, got no job 'cause they shut it down. Hear they got cheap stuff from China. Gonna get trainin' fer sumpthin', waitin' fer a word. Need to feed the wife and kids," he said.

I did manage to find him some temporary work.

I intend to give you the facts as I found them from many credible sources, which will be clearly identified. I won't pull any punches; if they do well, I will report the positive results fairly. However, you'll also get the other side of the coin if the program has issues that harm the country, cause confusion, set impossible expectations, mislead on climate change, or create unreasonable financial burdens.

Author's Note

Let me explain the structure of this book. I'll begin by introducing two concepts that I think are critically important to everyone - Systems Analysis and Cause and Effect. Then, I'll guide you through a discussion of the 20 most populous countries and theories about Earth's population, and include real, contributed articles with commentary from experts and me. You'll also find statistics and factual information about ice melting, the New Green Deal (NGD), and the actions being taken by responsible parties and corporations to mitigate the impacts of a runaway NGD. You might be amazed by the actions being taken outside government oversight, resulting in critical advances in carbon emission mitigation - ironically, from the industries targeted by the New Green Deal folks.

Believe me, there are a lot of critically vital facts to learn about throughout this manuscript. First, thank you for taking the responsible step of acquiring this book and showing an objective interest in understanding the issues and actions being taken on both sides of the fence, as they say.

The Buzz Bin – Tantalizing Tidbits, Oh Yeah, For Sure

The changes are disturbing and obviously misleading; they are not built on solid ground. In the third paragraph, McKibben claims that polling for the Green New Deal is "through the roof," when the link he provides puts approval at just 43.7 percent (Survey must have been conducted on another planet!). The Yale Program on Climate Change and Communication provided the foundation for reading this book, yet this survey raises questions about the veracity of what NGD proponents tell us. 97% of survey respondents admitted they knew "little or nothing at all" about the NGD. Amazing! This will be discussed in more detail under the "Embraced by the Uninformed" section later in this book.

WOWZA!

Is this the end of the New Green Deal? Is it all a big money grab or an attempt to take over our government? Have we been duped? I do believe in climate change, but how much do I really believe now?

The big reveal might be crawling out from under the rug; stay tuned!

I wonder what else is really going on. Is it time to face the truth? I believe so. And I suspect you will be left wondering what else was hiding under the rug.

Prologue

There is a lot of apprehension concerning the New Green Deal initiatives. This writing will assemble massive amounts of reliable information and available data to bring some semblance of order and understanding to know what has happened thus far and what we can expect. I assure you that there have been many dynamic changes and expectations from the time of the NGD's inception until 2025. You have the right to be fully informed of these expectations, changes, and your role in living within the New Green Deal (NGD) programs affecting our lives, both figuratively and financially. Strap in - it's an extraordinary ride.

The infamous Congressional Bill HR-109, also known as the New Green Deal, claims the world needs to cleanse all greenhouse gas emissions affecting climate change. It was released as an unfunded resolution bill, meaning it was not actually voted on by either house of Congress, with proponents choosing to vote "present" rather than in favor. Doesn't give me a high sense of confidence—how about you?

Everything was set in motion, and unreasonable expectations flourished. People without real understanding of the impact on world economies scrambled to embrace the concept in an insane, impossible time frame - completion by 2030 and 100% by 2050. Instantly, billions of dollars were appropriated to counter the threat of melting glaciers and polar ice caps, even while ignoring the science that nothing could be done about it. The damage had been done thousands of years ago through plate shifts and movements beyond our control. Yet, we still want to fund the whys and how's shows. Hey, the horse is already out of the barn! Let us focus on something we might actually be able to help with – let us get realistic!

Water flooding low-lying cities and countries, air quality so bad it could gag us into unknown sicknesses - it was even proposed we were on the brink of disaster, no, a growing catastrophe, unless we eliminated greenhouse gas emissions entirely. I quote a prominent figure saying, "Setting a goal only to reduce our emissions, but not eliminate them, won't do it." And guess what? We do not have a world buy-in. Did they forget to tell us that India, China, North Korea, and others were not going to participate in this charade? Yes, charade - because they are ignoring the issue altogether. The fact is, they need to take care of their massive populations with no possible mitigating alternatives.

Wait, I haven't dropped the other shoe yet!

It seems we don't have much "Dutch" in us - we can't just stick a thumb in the dike to hold back what disasters we perceive are befalling us. The real disaster is the people who have no idea how big a tiger by the tail they've got. It keeps growing, yet they continue to write unrealistic action plans costing billions of tax dollars - the latest released in 2024 - that are clearly impracticable to execute and an unreasonable and logically unattainable goal. And I do not just mean in the United States; they also include billions in funding for other countries, which I object to. Meanwhile, they talk and rave about what is causing distress in our country and worldwide, some of which there is no remedy that they fail to reveal. Many US and foreign voices - some in outright refusal or silence - clearly say "they don't want to play."

I assure you, many opinions are in opposition, and this is a far-reaching, horrendously costly RESOLUTION we are being forced to accept. I cannot include the new resolution in this book, but I can give you its name: "Delivering a Green New Deal: How the Bipartisan Infrastructure Law and the Inflation Reduction Act Bring the Green New Deal to Life," last updated in March 2024. I cannot include the whole 122-page document here, but I think it's worth reading to see what they want to do - and how you and I will have to pay for it. Hint: check out how much the states will have to contribute to this unreasonable goal. An especially crucial point here is that this bill and all the preceding NGD wants are not bills; they are resolutions that do not have a bill, and your representative or senator does not vote upon monies and is not controlled. SHOCKING! I discovered on 10/24 that the World Bank for NGD funds

CANNOT FIND 41 BILLION DOLLARS BASED ON AN AUDIT! I personally believe we should follow the expenditures and see who or what we can discover. Sounds shady to me.

I partially agree that rising temperatures, such as longer-lasting droughts, massive crop failures, and ancient and current populations having to move due to water issues, wildfires, and severe storms like hurricanes and floods, can all be attributed to climate change. An especially critical point I wish to make is that the Earth plays a significant role in what is occurring. There is considerably more that will be discussed.

Today, unusual events are becoming more commonplace, intense, and sometimes unpredictable. Comprehensive mitigation of carbon emissions is necessary in any plan of action. But how can this reasonably be done? We will discuss potential solutions, but they cannot be guaranteed to produce effective results due to various limitations. However, efforts are ongoing to find workable solutions.

Can we attribute some of these issues to forces beyond our control? The answer is that some of it may be out of our hands. Scientific evidence shows that Earth has issues beyond human control, and the Sun exacerbates these problems. Let me explain.

For example, did you know that the Earth has shifted and continues to shift, and that the polar regions are getting hotter? Fact: Believe it or not, there is proof that not all of this is caused by human activity. The solar system, the Earth itself - from its core to tectonic movement - and the Sun all play a role in this evolving saga, which has triggered a global spending frenzy to try and control the environment. I assure you that I will address these matters later since there is not one thing we can do to remediate the damage already done, but we can stop further damage if permitted.

I am starting to believe the world needs to step back, take a deep breath, and reevaluate how we are managing global warming. Polar ice melts, Earth's axis shifts, and rising land masses contribute to more movement and more changes. I hope you noticed I did not say we could solve the whole issue because some really are beyond our control. The question is, are we critically spending on the right things or just cherry-picking solutions? I do not think we are.

I have reviewed hundreds of documents written by authoritative government bodies, physicists, independent agencies, thoughtful individuals, and collegiate studies. While there is no absolutely conclusive solution to global warming, it is simply the largest elephant in the room screaming for what I think is not by any means an easy solution.

I want to clarify that I am not against acknowledging global issues. In fact, I appreciate the efforts to make the public aware of the dangers to Earth, such as through the New Green Deal (unfunded) program—real shock and awe, as they say.

I have evidence that the NGD advocates are using heavy-handed techniques to squash or eliminate entire industrial sectors to implement underperforming, costly, and dangerous so-called alternative solutions, all while promising huge job opportunities that will not - and cannot - materialize. Their solutions cannot absorb the millions of laid-off workers from the industries they are literally destroying. These realities will be fully exposed, I assure you. By the way, those industries are fighting back.

I am also acutely aware that many issues exist, and some can be reversed within the United States. Our businesses are taking serious steps; you will see these actions documented in this book to their credit. My family and I personally do whatever we can to help, but we also realize that we are but a small part of the overall picture. Everyone must contribute.

I believe that harmful emissions from citizens of the Earth are at fault for many unchecked emissions due to a number of contributing sources. However, the idea that the USA is the largest culprit is not wholly true—far from it.

However, I want our hard-earned bucks to be used only in the USA to assist in areas we know will have an effect. Some things our government does with our money raise concerns, such as wrongful political awards, like what happened with Solyndra or SunPower, which recently went bankrupt, affecting millions of solar panel products and installations. This cost the government all the upfront money. And, of course, this happened in California, which can ill afford another scandal, especially after a few billion dollars were spent on a fast train that never materialized!

Many so-called second—and third-world nations with massive populations—and a fair number of second-rate nations that conveniently look the other way—need major oversight. I might as well throw my opinion into the mix: supposedly, poor nations claim they lack the resources to act, but they can do some things. China, Russia, and many others seem to disregard the issue and give only lip service to the world organizations that monitor these problems.

In the meantime - and yes, I am being critical - the USA is proposing to throw 93 trillion dollars at the wind, supporting our nation and countries that either lack capability or do not want to bear the burden. Our US Treasury Secretary has not yet floated this massive sum without Congressional approval. This money is tied to another organization, which I will discuss, as it is linked to the New Green Deal in the United States and globally.

Belligerent While Spitting in The Face of The World - *Unconscionable!*

Here's the other shoe, and I believe it involves a secret, like-minded world cabal that met in Mexico in early 2019. While I haven't read all 110,000 **pages** (and I doubt I will), I think I have read enough to gather their intent. They are "one-worlders" looking for global equality. How? One of their goals is to heavily tax the wealthy and those of us in the middle class to support the world's poor. (I wonder who would administer this - surely not the ineffective UN!)

Now, let me tell you what I believe is happening. Putting two and two together, I saw that in the same year, our Treasury Secretary announced that she wanted $78 trillion of our tax dollars over an 8-year period. Her request included using our tax dollars to help the poor in other nations combat climate change. Coincidence? I think not. If you want to believe her, she probably has a New York bridge she wants to sell to you.

I am very worried about the NGD's ultimate goals. Right now, they cannot meet most of their suggested targets for 2030 or even 2050, and there are many pushbacks from industries, states, and commerce. They are lining up against the NGD because of its cumbersome requirements, overwhelming costs, and impracticality.

China, India, Germany, and North Korea—almost one-third of the world's most populated nations—are turning their backs on the world's new green initiatives! So why is the NGD pressuring us so hard when this is happening? All of our altruistic moves and spending will have at best, a minor effect.

I strongly believe that if we, in the United States and other like-minded people, clearly identify what we can have positive results in controlling the issues we can, we should address our issues as best we can and prepare for the future to support our populations.

Climate news release: China has been building a new coal-burning plant weekly for the past year and plans to continue doing so. India is doing the same and also constructing nuclear plants. Overwhelmed by the rising costs of oil and gas, Germany has given up and reopened its previously shut-down coal plants, with new ones underway. North Korea, too, is using and building coal plants. These actions are being taken to support population growth and the failure to generate enough electricity for their industrial expansion.

Here is a unique idea: I want Senator Markey and Congresswoman Cortez to get on a plane, visit the leaders of these countries, and insist they come to heel. We, the People, do not want to fund them anymore! We must fund our homeless veterans who bleed for us in many ways.

Author's Suggestion: After reading my extensive research, it has become clear that deploying the New Green Deal is like blindfolding a person and training them to shoot, hoping they would hit the target.

We never get updates on success, setbacks, or redirection based on outcomes, yet they continue asking for and receiving our tax dollars while other critical programs remain underfunded. Oh, and when we hear about failures - like the solar manufacturing disasters - it's only after the damage. My research shows they are missing the mark, are not transparent, and are uncontrollable. Industries are even rejecting some initiatives as too costly or impractical. Take the trucking, auto, and aircraft sectors, for example. However, promising technological advancements are being developed within the mandated timeframes, producing surprising and welcome results.

The industry is protecting its interests and stepping up, as I anticipated. Through Research and Development, private companies are meeting or even surpassing compliance goals ahead of schedule.

Key To This Major Situation: Moving forward, I propose and highly recommend that a blue-ribbon oversight commission, composed of experts in the relevant sector and open-minded individuals with proven records, take a global view of the US (NGD) to prioritize and fund areas that can be successfully deployed and realized. The government does not have a stellar reputation for successful completion and spending, but perhaps this could be curtailed, allowing real progress to be made. Failure has its limits, and political favoritism isn't always successful.

A comprehensive roadmap for all desired projects should be prioritized without direct government intervention (but using normal state-funded tax dollars first). This approach is fair to all of us who will ultimately pay the bill. The roadmap should then be presented for an authorizing vote, with all expectations clearly defined, including metrics.

Overall, systems analysis should be conducted to achieve this, not the current haphazard approach, which wants to throw gooey stuff against the wall to see what sticks. There are too many priorities and too much money at stake to address everything at once.

I also call upon the UN, to whom we contribute space and a large amount of funds each year, to grow some power and take this on outside the United States. If necessary, they should sanction or even remove those nations that defy their intent to tackle and remediate this global issue - it's a real, one-world issue.

Before delving into more of the book's content, I wish to present a concept that I think is extremely important:

Systems Analysis Knowledge

As a business studies professor, I maintain that systems analysis is a critical subject every student should study. It provides essential knowledge of both simple and complex systems, offering a global view of a complete system - a roadmap.

I believe all students should be required to take Systems Analysis and CIVICS courses from mid-high school through higher education that teach the reality of how things work, from start to desired completion, addressing all the complexities needed for effective societal outcomes. I believe this education would significantly benefit both the students and society in general. My desired effect, after each class (after the bemoaning of every student was suffered), is that their class evaluations consistently gave me 4.0 remarks, with students stating their lives had changed and that they now looked at their surrounding world in an enlightened way. Over one hundred students are wandering about now with impressive knowledge of how to critically look at a project with wide-open eyes and brains in gear. It is a grounding experience and perhaps makes life more realistic.

Another valuable concept is *Cause and Effect,* and I want to introduce you to a definition of an ancient and well-worn principle.

Cause and Effect

Definition:

The Law of Cause and Effect is like a rule that says every event that happens is the result of a specific cause. Imagine you have a row of dominoes; if you knock the first one over (cause), the rest will fall down in sequence (effect). This rule helps us figure out why things happen and what could happen next.

Here is another way to think about it: The Law of Cause and Effect is like a chain of events. Every link in the chain is connected. If one link moves (cause), it can pull the next one (effect), and so on until the end of the chain. When we talk or write about things logically, we use this law to show how each part of our proposals can connect, just like the links in a chain. If the links do not fit together well, our proposal might not be strong.

To use this law, we often ask questions like, "Why did this happen?" or "What will happen if I do this?" By doing so, we can hopefully make better choices by thinking ahead about the effects of our actions or understand events better by looking at their causes.

Let me explain further by giving you, the reader, some clear examples of this concept in practice. As you will see, I am applying this principle to what I will call the not-clearly defined, seemingly uncontrolled concept of the **New Green Deal** and its approach to resolving national and global issues.

I should inform you that I do not disagree with the conservation concept. However, I do not fully agree with the applied and radical approach the United States is taking, seemingly without proper controls or reasonable expectations in place. To me, it seems like an unreasonable expectation of success, with mixed and often unacceptable results for our population. Of course, we're talking about clearly defined, plausible solutions for expenditures - solutions that yield measurable, desired results for our scarce tax dollars.

Newton's third law of motion, first formulated by Sir Isaac Newton in his book *Mathematical Principles of Natural Philosophy* in 1687, coined the phrase: "For every action, there is an equal and opposite reaction." While this may sound complex, it essentially means that cause and effect apply to almost every aspect of life. For each action, outcomes might not be random but determined by the factors that led up to them.

Cause and effect describe the relationship between events or actions, where one event (the cause) leads to another event (the effect). It is the concept that every action or event has a consequence, and this can potentially snowball, causing other cascading results and undesired issues.

Cause:

A cause refers to the event, action, or condition that directly results in another event or action, known as the effect. A cause can be a single action or a series of events that lead to a particular outcome. It is the reason behind the effect, and without it, the effect might not occur. Understanding the cause of an event helps us comprehend the factors that led up to it, allowing us to somewhat predict and potentially prevent similar events in the future.

Effect:

The effect refers to the event, action, or condition that is directly or indirectly caused by another event or action, known as the cause. It is the outcome or consequence of the cause, and it might not occur without it.

The effect can be positive, negative, or neutral, depending on the circumstances and the specific cause that led to it. Understanding the effect of a cause helps us better understand the impact and significance of the cause, allowing us to evaluate the consequences of different actions or events.

Understanding cause and effect together allows us to better grasp the relationships between events and phenomena. This concept can be applied to a wide range of fields, including science, psychology, social studies, and even in our approach to the **New Green Deal**, which, to some extent, involves speculative outcomes that may not be entirely within our control.

Cause and Effect Meaning

We have already covered the definition of cause and effect, but what does cause, and effect mean in its simplest form? Essentially, the cause is 'why' something happens, and the effect is the 'outcome' of what happened.

In the context of the English language, 'cause' can be both a verb and a noun.

As a verb, 'cause' means to make something happen or to be responsible for a particular outcome. For example, "Smoking cigarettes can cause lung cancer" means that smoking can lead to lung cancer, the end effect.

As a noun, 'cause' refers to 'the reason' why something happens. For example, "The cause of the accident was a failure to obey traffic signals." In this context, 'cause' is a noun that describes the reason or event that led to the accident.

'Effect' is a noun in the context of cause and effect. It refers to the outcome, consequence, or result produced by a particular cause or action. For example, "The effect of the medication was to reduce pain and inflammation." In this context, 'effect' is a noun that describes the outcome or result of taking the medication.

Examples of Cause and Effect in Sentences

- Heavy rain caused the streets to flood.
  Cause: It rained heavily. **Effect**: The streets flooded.
- She had to book another ticket after missing her flight.
  Cause: She **missed** her flight. **Effect**: She had to book another ticket.
- The team's daily practice contributed to their championship win.
  Cause: The team practiced every day. **Effect**: They won the championship.

Examples of Cause and Effect in the Real World

There are numerous examples of cause and effect throughout history intertwined in historical events, nature, and the day-to-day lives of individuals. Let's take a look at some examples:

Historical Events

- **Cause:** The Industrial Revolution and urbanization.

  Effect: The growth of labor movements and demands for better working conditions and labor laws.

- **Cause:** The Civil Rights Movement and nonviolent protests.

  Effect: The passage of the Civil Rights Act and Voting Rights Act in the United States.

- **Cause:** The fall of the Roman Empire.

  Effect: The rise of feudalism and the Middle Ages in Europe.

- **Cause:** The invention of the printing press.

  Effect: The spread of knowledge and literacy, leading to the Reformation in Europe.

- **Cause:** The Great Depression and economic inequality.

  Effect: The rise of socialism and the welfare state in many countries around the world.

Nature

- **Cause:** Deforestation and habitat destruction.

  Effect: Loss of biodiversity and increased risk of species extinction.

- **Cause:** Natural disasters such as hurricanes, earthquakes, and volcanic eruptions.

  Effect: Damage to infrastructure and property, displacement of people, and loss of life.

- **Cause:** Drought and heat waves.

  Effect: Crop failures, food shortages, and damage to natural landscapes.

- **Cause:** Soil erosion and depletion.

  Effect: Loss of soil fertility, decreased agricultural productivity, and increased risk of landslides and floods.

Lives of Individuals

- **Cause:** Regular exercise and a healthy diet.

  Effect: Improved physical fitness, overall health, and reduced risk of chronic diseases.

- **Cause:** Hard work and dedication.

  Effect: Achieving personal and professional goals, career advancement, and financial stability.

- **Cause:** Quality education and access to resources.

  Effect: Increased opportunities for personal and professional growth, better job prospects, and higher income.

- **Cause:** Environmental conservation and sustainable practices.

  Effect: Preservation of natural resources, protection of wildlife and ecosystems, and mitigation of climate change effects.

- **Cause:** Community involvement and volunteer work.

  Effect: Building stronger communities, creating positive change, and helping those in need.

- **Cause:** Innovation and creativity.

  Effect: Advancement in science, technology, and the arts, leading to improved quality of life and new opportunities.

Sayings Related to Cause and Effect

- "You reap what you sow." This means that your actions will have consequences, and you will experience the effects of those actions.

- "Actions speak louder than words." This means that what you do is more important than what you say, and your actions will have a greater impact than your words.

- "What goes around comes around." This means that the consequences of your actions might eventually return to you, either positively or negatively, depending on what you have done.

Conclusion

The cause of you reading this material was probably a desire to learn more about cause and effect. Hopefully, the effect is that you have learned more about it!

But seriously, cause and effect is an important concept that helps us understand the relationship between events and their outcomes. Whether in personal, historical, scientific, or natural contexts, the concept of cause and effect helps us analyze the factors that contribute to outcomes and can help us make better, decently marginal effect decisions, even in situations that may seem unsuccessful.

This book's intent is to critically review the New Green Deal and its actual effectiveness. Many initiatives, both taken and proposed, will be reviewed for their reasonable successes, programs that are marginally successful, and those where damage seems already beyond hope of address.

My purpose for this book's content is to research and deliver empirical information concerning the methods and activities of the tenets of HR-109, the New Green Deal (NGD). I will examine its approach and how its actions have impacted various sectors, including the economy, environmental focus, manufacturing, health, automotive, aerospace, transportation, sea transport, wind and solar energy, the fossil fuels industry, trucking, polar issues, earth core movement, the Sun's effects on Earth's temperature, the current and future costs of the NGD, and its effects on autos, boats/ships, trucking, labor, and many other areas. This will be presented in a Pro/Con format to offer a professional viewpoint.

The NGD does not have full approval from the majority of the US population, and its goals have negatively affected a large sector of the world due to its debilitating costs, which divert critical funding away from necessary system sectors and people who rely on government support for these programs.

Really, Just What Is Climate Change?

Climate change refers to a big, long-lasting shift in global weather patterns. It is commonly known as global warming because the Sun warms our atmosphere. It is called global warming because the Sun warms our atmosphere. Global warming because the Sun warms our atmosphere Global warming because the Sun warms our atmosphere, and the Earth traps those warming rays.

Sounds simple. Well, it is not by any means. The world's populations seem to be complicating this natural balance by introducing a layer of greenhouse gases into the atmosphere. One of these is carbon emissions, but it's important to understand that this buildup has been happening for an incredibly long time - it is not a recent development and predates human activity. I am drawing information from World Heart Worldwide, and while I believe they correctly state that the blanket of emissions is causing the oceans to warm, which leads to more significant storms like hurricanes and typhoons, the melting of polar ice started thousands of years ago. However, this is not totally the fault of carbon; it also has to do with Earth's shift and continental rise, which I will address a bit later.

Since the industrial age began 250 years ago, the planet has warmed by 1.1 degrees Celsius, or 1.9 degrees Fahrenheit. According to the International Panel on Climate Change, this could rise to a projected 7.2 degrees Fahrenheit by the year 2100. They point to fossil fuels such as coal, oil, and gas as the major contributors.

I do not deny that the industrial period we are in has contributed, but the good news is that there are responsible companies and organizations bringing forth advanced technology. I am not talking about millions of acres of solar panels on every home or thousands of square miles of wind generators, which contribute only a small portion to the national grid. I will elaborate on this later.

While we think mainly about our circumstances, I want to very clearly tell you there are major countries like India, China, Russia, North Korea, and an exceptionally large list of poorer countries that simply do not care or afford to participate. These countries represent over half of the world's population and land mass and struggle - how does anyone control them? They are belligerent at best, desire or they just do not have the money.

Here are some profoundly serious facts from current census data that will stun you. While India's population growth has slowed remarkably over the last few years, it is still growing faster than China's and is expected to surpass China in population by 2026, when both will have about 1.46 billion people. After 2030, India is expected to be the most populous country in the world, reaching a peak population of 1.65 billion by 2060.

Rank	Country	Population
1	India	1,452,453,091
2	China	1,418,940,938
3	USA	345,643,917
4	Indonesia	283,750,127
5	Pakistan	251,731,364
6	Nigeria	233,245,475
7	Brazil	212,094,289
8	Bangladesh	173,811,315
9	Russia	144,723,218
10	Ethiopia	132,457,260

11	Mexico	130,988,470
12	Japan	123,676,338
13	Egypt	116,752,109
14	Philippines	115,954,402
15	D.R. Congo	109,689,328
16	Vietnam	101,059,458
17	Iran	91,667,462
18	Türkiye	87,498,710
19	Germany	84,495,886
20	Thailand	71,662,337

Source: Worldometer live calculator as of 8/12/24

How many people can the Earth sustain? Opinions vary widely; I am afraid.

The world population has been expanding non-stop for six hundred years and is expected to continue growing for at least one hundred years, reaching more than 11 billion people by 2100. There are also straightforward signs that Earth cannot cope with the growth of the human population. We alter carrying capacity when we manipulate resources in a natural environment. If a population exceeds carrying capacity, the ecosystem may become unsuitable for the species to survive. If the population exceeds the carrying capacity for a lengthy period of time, resources may be completely depleted. If the global fertility rate does indeed reach a replacement level by the end of the century, then the human population will stabilize between nine billion and ten billion. As far as Earth's capacity is concerned, we will have gone about as far as we can go but no farther. More people mean an increased demand for food, water, housing, energy, healthcare, transportation, and more. And all that consumption contributes to ecological degradation, increased conflicts, and a higher risk of large-scale disasters like pandemics.

There are limits to the life-sustaining resources Earth can provide us. In other words, there is a carrying capacity for human life on our planet. Carrying capacity is the maximum number of species an environment can support indefinitely. Every species has a carrying capacity. I personally have been to Vietnam, Russia, Europe, Brazil, Mexico, the USA, Japan, Hong Kong, Taiwan, and the Philippines and have seen the effects of population growth; I fully understand the issue.

Let's Get Real How Many People Can Earth Handle?

Tokyo is the world's most populous city, with 37 million people, 60,000 restaurants, and 167 skyscrapers.

Toward the end of 2022, the human population on Earth is expected to reach eight billion. To mark the occasion, BBC Future takes a look at one of the most controversial issues of our time. Are there too many of us? Or is this the wrong question?

One moment, the valley was a tranquil, swampy wetland. Grasses and palm trees cast fuzzy shadows on the water below. Fish lurked warily at the edges of mangroves. Orangutans sought out fruit with leathery fingers. Then, a dormant giant awoke from its sleep.

It was around 72,000 BC on the island of Sumatra, Indonesia. The Toba super volcano erupted in what is thought to be the greatest eruption event in the last 100,000 years. A series of thunderous explosions blasted out 9.5 quadrillion kilograms of ash, which billowed out in sky-darkening clouds that crept around 47 km (29 miles) into the atmosphere. (Being unrealistic and a smart butt, I could suggest that the NGD folks look into getting funding for volcano plugs to stop this; they funded everything, didn't they?)

In the aftermath, a vast area across Asia was blanketed in a layer of soft dust 3-10 cm thick. It choked water sources and stuck to vegetation like cement – deposits from Toba have been found as far away as East Africa, 7,300 km (4,536 miles) west of the eruption. But crucially, some scientists believe it plunged the world into a volcanic winter that lasted decades and nearly made our species extinct.

In 1993, a team of American researchers studied the human genome for clues to its deep past and discovered the tell-tale signature of a major 'population bottleneck' – a moment when humanity shrank so drastically that all subsequent generations outside Africa were significantly more closely related.

Later studies have revealed that in this precarious era, which may have occurred between 50,000 and 100,000 years ago, our collective numbers may have hit as few as 10,000 people – equivalent to the current population of the sleepy settlement of Elkhorn in Wisconsin or the number who attended a single drive-through wedding in Malaysia in 2020. The least affected part of the world was Africa, where genetic diversity remains high to this day – on this single continent, there are larger genetic differences between certain groups than there are between Africans and Europeans.

Some think this timing is not a coincidence – they believe the Toba volcanic eruption did it. The idea is hotly disputed, but there is no doubt that much of humanity is descended from a relatively modest number of super-hardy ancestors. At times, the inhabitants of entire regions of the world have been in great peril.

Fast-forward 74,000 years, and the human population has undergone a population explosion, colonizing nearly every habitat on the planet and exerting our influence on even the remotest corners - in 2018, scientists found a plastic bag 10,898 m (35,754 ft) below the ocean's surface at the bottom of the Mariana Trench, while another team recently discovered man-made 'forever chemicals' on Mount Everest. No part of the world seems pristine – every lake, forest, and canyon seems to have been touched by human activity.

At the same time, our sheer numbers and ingenuity have enabled humanity to achieve feats that no other animal could dream of – splitting atoms, sending complex equipment nearly a million miles (1.6 million km) to observe planets forming in distant galaxies, and contributing to a staggering diversity of art and culture. We collectively take 4.1 billion photographs daily and exchange between 80 and 127 trillion words.

On the uncanny specific date of 15 November 2022, the United Nations has predicted that eight billion humans will be alive at the same time, up to 800,000 times as many as there were survivors of the Toba catastrophe.

Today, our population is enormous, with such little genetic diversity—outside Africa—that one researcher recently observed it is actually not that surprising that some people look uncannily similar to perfect strangers. A limited gene pool is constantly being recycled, and around 370,000 new opportunities (in the form of babies born) for these genetic coincidences to occur each day.

But with humanity's expanding population has come great division. Some view our rising numbers as an unprecedented success story—in fact, there's an emerging school of thought that we actually need more people. In 2018, the tech billionaire Jeff Bezos predicted a future in which our population will reach a new decimal milestone—a trillion humans scattered across our Solar System—and announced that he's planning ways to achieve it.

Others, meanwhile - including the British broadcaster and natural historian Sir David Attenborough – have labeled our swarming masses a "plague on the Earth." In this view, nearly every environmental problem we're currently facing, from climate change to biodiversity loss, water stresses, and conflicts over land, can be traced back to our rampant reproduction over the last few centuries. Back in 1994, when the global population was a mere 5.5 billion, a team of researchers from Stanford University in California calculated that the ideal size of our species would be between 1.5 and 2 billion people.

So, is the world currently overpopulated? And what might the future hold for humanity's global dominance? The debate over the ideal number of people on the planet is as fractured and emotionally charged as ever, but time is running out to decide which is the best direction.

An Ancient Concern

It was the late 1980s in central Iraq, and a team of archaeologists from the University of Baghdad was excavating a ruined library in the ancient city of Sippar. Amid the sand, dust, and antique walls, they found 400 clay tablets – records that had been lying forgotten in their tomb for over three and a half thousand years, still on the

24

same shelves they had once been added to by Babylonian hands.

But four in particular were special for them. They contained missing sections of a story found in fragments on separate tablets scattered across Mesopotamia, which still intrigues historians today.

"Twelve hundred years had not yet passed [since the creation of mankind], when the land extended, and the people multiplied…" goes the Atra-Hasis – the poem stamped into the clay by an anonymous scribe around the 17th Century BC. It's the Mesopotamian version of the Great Flood story found in many forms in cultures across the globe. It might contain one of the earliest mentions of overpopulation in the historical record.

Around the time the Atra-Hasis was written, the global population is estimated to have been between 27 and 50 million people, equivalent to the number who currently inhabit Cameroon or South Korea, or 0.3 - 0.6% of the total who call the planet home today.

During the millennium that followed, scholars appear to have fallen relatively quiet about any population concerns. Then, in Ancient Greece, they began to ponder the issue again.

The philosopher Plato had some strong views. After a period of rapid growth, in which Athens' population doubled, he lamented, "What is left now is like the skeleton of a body wasted by disease; the rich soil has been carried off, and only the bare framework of the district is left." Not only did he believe in strict population control managed by the state, but he eventually concluded that the ideal city should have no more than 5,040 citizens and thought setting up colonies was an effective way of offloading any excess. He also felt the importance of moderating consumption.

In Plato's magnum opus, The Republic, written around 375 BC, he describes two imaginary city-states – administrative regions governed almost like small countries. One is healthy and "luxurious" and "feverish." In the latter, the population spends and devours excessively, surrendering themselves to consumerism until they have "overstepped the limit of their necessities." Alas, this morally decrepit city-state eventually resorts to seizing neighboring lands, which naturally spirals into war – it simply can't sustain its large, greedy population without extra resources.

Plato had hit upon a debate that's still raging today: is the human population the issue, or is it the resources it consumes?

It took more than five centuries after Plato before the global scale of our population explosion became clear. The author Tertullian, who lived in the Roman city of Carthage, pre-empted modern-day observations about our destructive multitudes by a long shot. In 200 AD, when the total human population had reached 190-256 million - somewhere around the number who currently inhabit Nigeria or Indonesia – he believed the entire world had already been explored. People had become a burden on the planet.

Over the next 1,500 years, the global human population more than tripled. Eventually, this detached concern turned into panic.

Enter Thomas Malthus, an English cleric with a penchant for pessimism. In his famous work, An Essay on the Principle of Population, published in 1798, he began with two observations— that all people need to eat and that they like to have sex. When taken to their conclusion, he stated that these simple facts would eventually lead humanity's demands to outstrip the planet's supplies.

"Population, when unchecked, increases in a geometrical ratio. Subsistence increases only in an arithmetical ratio. A slight acquaintance with numbers will show the immensity of the first power compared to the second," wrote Malthus. In other words, large numbers of people lead to even larger numbers of offspring, in a kind of positive feedback loop – but our ability to produce food doesn't necessarily accelerate similarly.

These simple words immediately ignited enthusiastic fear in some and anger in others, which would continue to reverberate across society for decades. The former thought something had to be done to stop our numbers from spiraling out of control. The latter thought that limiting the number of people was absurd or unethical and that every effort should be made to increase the food supply.

The fewer-people camp was particularly critical of the English Poor Laws introduced hundreds of years earlier. These laws involved payments to those living in poverty to help them care for their children. It was speculated that these laws encouraged people to have larger families.

At the time Malthus' essay was published, there were 800 million people on the planet. It wasn't until 1968, however, that modern concerns about global overpopulation emerged when a professor at Stanford University, Paul Ehrlich, and his wife, Anne Ehrlich, co-authored The Population Bomb. The Indian city of Delhi was the inspiration. They were returning to their hotel in a taxi one evening and went through a slum, overwhelmed by human activity on the streets. They wrote about the experience in a way that has been heavily criticized – especially since London's population at the time was more than double that of Delhi.

The couple authored their book because they were concerned about mass starvation that they believed was coming, particularly to those in developing countries, but also in places like America, where people were starting to notice the impact they were having on the environment. The work has been widely credited or accused of, depending on the view—triggering many of the current anxieties about overpopulation.

Of course, discussions about how many people there should be have never been purely academic. At times, they have been hijacked to justify persecution, ethnic cleansing, and genocide. In each case, the perpetrators have been intent on lessening the populations of specific groups of people, such as those from a certain social class, religion, or ethnicity (rather than humanity as a whole). Still, they are sometimes seen as examples of the dangers that the very concept of overpopulation can pose.

As early as 1834, just three and a half decades after Malthus' essay was published, the English Poor Laws were scrapped and replaced with stricter ones. This was partly over Malthusian concerns that this social class (who he called "peasants") was reproducing too much and had the result of forcing orphaned children into bleak, unsanitary workhouses like the one depicted in Charles Dickens's novel Oliver Twist.

Over the coming centuries, eugenics was continually disguised as population control – or received support from the movement – such as during the forced sterilizations of people from minority ethnic groups in 1970s America. It was also used to curtail individual freedoms. In 1980, China introduced its controversial one-child policy, which was widely seen as an invasive violation of the population's sexual and reproductive rights.

A Contentious Future

As a result of its controversial history, population engineering is a deeply divisive area.

Today, any policies involving quotas or targets to increase or decrease the human population are almost universally condemned, except by a handful of extremist organizations. The risk of these incentives leading to coercion or other atrocities is seen as too great. But there is little agreement beyond this.

At one end of the spectrum lie those who see lower fertility rates in some areas as a crisis. One demographer is so concerned by the localized drop in the birth rate in the UK that he has suggested taxing the childless. As of 2019, there were 1.65 children born in the country per woman on average. This is below the replacement level – the number of births required to maintain the same population size – of 2.075. However, the population grew overall because of migration from other countries.

The opposing view is that slowing and eventually halting global population growth is not only eminently manageable and desirable but can also be achieved via entirely voluntary means—methods such as simply providing contraception to those who would like it and educating women. In this way, proponents of this position believe we could benefit the planet and improve the quality of life experienced by the poorest citizens worldwide.

One organization that believes in this approach is the UK-based charity Population Matters, which campaigns to achieve a sustainable global population. They advocate for addressing the pressures consumerism, particularly by the developed world, is placing on the planet, while highlighting the role that population size has to play. To this end, they urge people to take individual responsibility for the environmental crisis and work toward ending global poverty and inequality through debt relief and overseas aid.

"We deplore any form of population control or coercion, restriction of choice," says the director, Robin Maynard. "So, it's about enabling access, enabling choice, and meeting rights. That is actually the most effective way that people will ultimately make decisions that are good for them and the planet."

On the other hand, others advocate for shifting the focus away from adjusting the number of people in the world, however softly or indirectly this is achieved, to our activities. Supporters argue that the quantity of resources each person uses up has more of an impact on our collective influence and point out that consumption is significantly higher in wealthier countries with lower birth rates. Reducing our individual demands on the planet could shrink humanity's footprint without stifling growth in poorer countries.

In fact, Western interest in curtailing population growth in less developed parts of the world has been accused of having racist undertones.

Finally, there's the fatalistic 'solution' to the perennial population question: do nothing. This view relies on the highly unstable dynamics of our global population – it's set to grow significantly, but then it will shrink. Every camp may get what it wants in the end, though not forever.

Estimates vary, but we're expected to reach 'peak human' population around 2070 or 2080, at which point there will be between 9.4 billion and 10.4 billion people on the planet. It may be a slow process - if we reach 10.4 billion, the UN expects the population to remain at this level for two decades. But eventually, after this, the population is projected to decline.

In the book 'Empty Planet: The Shock of Global Population Decline", the authors present a vastly different vision of the future from the one we're used to, in which the world grapples with the new challenges and opportunities that depopulation could present.

Amid all the controversy and uncertainty, it can be hard to know what to think. Here's how the number of people on the planet might affect a few key aspects of our lives in the future: the environment, economy, and our collective well-being.

An Environmental Challenge

With bumpy handheld shots, the camera stalks through the Madagascan Forest. It's thick with trees and the thrilling mystery of an unknown wilderness. Then suddenly, there it is: a white blur bounces across the frame and off into the distance. The animal is a sifaka - a shy and elusive lemur with long limbs, pale fur, and a black face, like a lanky teddy bear.

The brief encounter is part of a BBC documentary, Indian Ocean with Simon Reeve, and the presenter soon reveals a startling caveat to the team's lucky find. This is not the wild after all – it's the Berenty reserve in Southern Madagascar, a tiny patch of forest surrounded on all sides by commercial plantations and one of the last remaining places this rare creature calls home.

At the visitor center, Reeve says he's been reliably informed that nearly all wildlife camera people who film in the area set their equipment up in this location, where the lemurs are most abundant, facing away into the forest so as not to capture any buildings. While viewers may think they're catching a glimpse into the wild unknown, you could argue that what they're really getting is a carefully curated illusion of untamed nature.

The documentary highlights the 'wilderness myth' – a misunderstanding that can occur when people are presented with majestic footage of the natural world that excludes humans entirely or drastically downplays our presence, suggesting that there are still vast tracts of untouched land.

Satellite images are a particularly powerful tool for shattering this notion: from the air, many countries are revealed to be heavily adapted for human use. As far as the eye can see, the land is a patchwork of agricultural fields, threaded with roads and row upon row of buildings. Some landscapes have been so transformed in just a few decades by engineering works or deforestation that they are barely recognizable.

These changes come with some startling statistics. According to the UN Food and Agriculture Organization (FAO), 38% of the planet's land surface is used to grow food and other products (such as fuel) for humans or their livestock – five billion hectares (12.4 billion acres) in total.

And though our ancestors might have lived among giants, hunting mammoths, mega-wombats, and 450 kg (1,000 lb.) elephant birds, today we are the dominant vertebrate species on land – a grouping that includes everything from geckos to lemurs. Humans account for 32% of terrestrial vertebrates by weight, while wild animals account for just 1%. Livestock accounts for the rest.

The wilderness preservation charity WWF has found that wildlife populations declined by two-thirds between 1970 and 2020, while the global population more than doubled

In fact, as our dominance increases, many environmental changes have been occurring in parallel, and several prominent environmentalists, from the primatologist Jane Goodall, famous for her study of chimpanzees, to the naturalist and TV presenter Chris Packham, have voiced their concerns. In 2013, Sir David Attenborough explained his views in an interview with the Radio Times: "All our environmental problems become easier to solve with fewer people, and harder – and ultimately impossible – to solve with ever more people."

For some people, alarm over humanity's environmental footprint has led them to decide to have fewer or no children themselves. This includes the Duke and Duchess of Sussex, who announced in 2019 that they would have no more than two for the sake of the planet. The same year, Miley Cyrus declared that she wouldn't have children because she thought the Earth was "angry."

A growing number of women are joining the anti-natalist movement and going on a "Birth Strike" until the current climate emergency and extinction crisis are dealt with. The trend was buoyed up by research from 2017, which calculated that simply having one fewer child in the developed world could reduce a person's annual carbon emissions by 58.6 tonnes of CO_2 equivalent (CO_2e) – more than 24 times the savings from not having a car.

One 2019 study, led by Jennifer Sciubba, an associate professor of International Studies at Rhodes College, Tennessee, analyzed the levels of population growth in over 1,000 regions in 22 European countries between 1990 and 2006 and compared them to how patterns of urban land use and carbon dioxide emissions changed over the same period. The team concluded that the sheer number of people had a "considerable effect" on these environmental parameters in Western Europe. Still, these were not the key factors in Eastern Europe.

This nuanced support for the idea that population growth leads to environmental degradation is backed up by many other studies, but so is the impact of spiraling demand for natural resources, especially in wealthier countries. In fact, many environmentalists now believe that the problems we're currently facing are largely due to consumption rather than overpopulation. From this perspective, concerns about the latter unfairly shift the blame onto poorer countries.

In 2021, a study found that population growth and non-renewable energy sources are degrading the environment in the US, while another found that economic growth and the use of natural resources in China from 1980 to 2017 led to higher carbon dioxide emissions.

Interestingly, in a call to the Department of Agriculture, I was told that in 2060 it was projected that the US population would be 360 million and that would be the max our agriculture output could sustain.

Intriguingly, other research has found that while the use of natural resources and urbanization in China both increase the rate of environmental destruction, these are partly offset by the availability of 'human capital'—essentially, the knowledge and skills of the human population.

Today, it is widely accepted that people are putting an unsustainable strain on the world's finite resources, a phenomenon highlighted by 'Earth Overshoot Day,' the date each year when humanity is estimated to have used up all the biological resources that the planet can sustainably regenerate. In 2010, it fell on August 8. This year, it was 28 July.

Whether the problem is too many humans, the resources we use up, or both, "I can't even fathom how more humans could be better for the environment," says Sciubba, who authored the book '8 Billion and Counting: How Sex, Death, and Migration Shape our World.' She suggests one way to make the case might be to view humans and the environment as the same entity, "but boy, that's a real hard argument to make."

However, Sciubba is keen to point out that the idea of an impending 'population bomb' coming to destroy the planet, as suggested by the Ehrlichs' book, is outdated. "At that time [when it was written], I think there were 127 countries in the world where women, on average, had five or more children in their lifetime," she says. In that era, population trends really did look as though they were exponential, and she suggests that this instilled a population panic in certain generations that are still alive today.

"But today there are only 8 [countries with fertility rates above five]," says Sciubba. "So, I think it's important for us to realize that those trends changed."

An Economic Opportunity

In 2012, the Singaporean government devised an unusual way for citizens to celebrate its independence and disseminated the instructions via a new rap song. The hit was intended to encourage young couples to have more children by mixing colorful innuendos with patriotic references to the country's culture and sights.

"...Let's make a lil' human that looks like you and me, exploring' your body like a night safari, I'm a patriotic husband, you a patriotic wife, Let; I'm a patriotic husband, you a patriotic wife. Let me come into your camp and manufacture a life...," went some of the lyrics.

It was released amid fears about Singapore's super-low fertility rate of just 1.1 births per woman as of 2020. This is an extreme example of what has become a common trend in wealthy countries, where people marry later and choose to have fewer children. In Singapore, this triggered concerns about the consequences for the nation's economy, leading the government there to call on its citizens to do their bit.

Population growth is a key concept in economics: the more people you have, the more goods or services you can produce and the more you can consume. Thus, population growth is the best friend of economic growth.

Author note: This is not the same model for all countries.

This is one reason that concerns about the growing population in developing parts of the world are sometimes seen as problematic – many developed countries are already densely populated, and this is partly how they achieved their wealth. Denying other countries this opportunity is seen as unfair and even racist.

However, slower population growth isn't always followed by an economic downfall. Take Japan, which pre-empted global trends in wealthy nations and achieved below-replacement level fertility rates as early as 1966, when it suddenly dropped from around 2 to 1.6. "I don't think it's the case that Japan's economy has declined to the extent that people sometimes portray it if you look at standards of living," says Andrew Mason, emeritus professor in the Department of Economics at the University of Hawaii. "They've invested a lot in human capital – so they have fewer children, but they've emphasized education and have very good healthcare systems."

Mason also points out that saving and investment are common in Japan: "So there's been increases in [monetary] capital and higher productivity as well. If you combine those things... Japan is a good case study in why not to panic [about declining birth rates]."

There are other ways of growing a country's economy. Mason points out that immigration often provides a helpful source for new workers without adding extra people to humanity's total. However, immigration also remains a contentious and highly politicized subject in some countries, so some countries won't have this option without cultural shifts in how it's perceived.

"Think particularly of countries like Japan and South Korea where they have [historically] been very resistant to immigration; they're going to find it more and more to their advantage [to change this policy]," says Mason.

Equally, the advantages that immigration can provide are inherently uneven - one country boosts its economy at the expense of another whose workers left.

There's a growing feeling that the global obsession with chasing economic growth at all costs is outdated and should be abandoned altogether. "One of the things that frustrates me about the overpopulation debate is a lot of comments come out of the same people's mouths – that we don't want there to be too many people, and we also want to make sure that the economy is always growing," says Sciubba. "In a world where there are fewer people, we really need a complete mindset shift away from growth equals progress," she says.

A Happier Future

However, demographics influence the environment and economy—they're also a powerful hidden force shaping the quality of people's lives worldwide.

According to Alex Ezeh, a professor of Global Health at Drexel University, Pennsylvania, the absolute number of people in a country is not the most crucial factor. Instead, she thinks that the rate of its population growth or decline is key to a country's future prospects, this determines how quickly things are changing.

Take Africa, where Ezeh explains that radically different rates of population growth are currently occurring, depending on where you look.

"In several countries, particularly in Southern Africa [one of five regions defined by the United Nations], fertility rates have really dropped, and contraceptive use is up – the rate of population growth slowing down, which is good news," says Ezeh.

At the same time, some Central African countries still have high rates of population growth due to high fertility and longer lifespans. In some places, it's well above 2.5% per annum, "which is massive," says Ezeh. "The population will double every 20-plus years in several countries."

Different countries can be on surprisingly different paths even within a single region. Ezeh gives the example of the East African neighbors, Burundi and Rwanda. While the former still has high growth rates—at 5.3 births per woman—in the latter, growth is slowing down, with 3.9 births per woman in 2020 compared to 4.5 in 2010.

"I think the conversation about size and numbers is misplaced," says Ezeh. "Think of a city that is doubling every 10 years – and that's several cities in Africa. Which government really has the resources to improve every infrastructure that currently exists every 10 years to maintain the correct level of coverage of those services?"

Ezeh explains that, in particular, it's difficult to support the development of human capital under conditions of extreme growth, which research has found plays a significant role in people's happiness in cities, even more than the amount of money they're earning. It's also considered an important predictor of economic growth, in addition to the number of people in a country.

"When economists think about it, a large population is great for many different outcomes, but do you achieve that large population in 10 years, 100 years, or 1,000 years? The longer it takes to get there, you can put in place the right structures in the system that will support that population," says Ezeh.

One factor with a well-documented role in slowing down this rate of growth is the education of women, which has the side effect of increasing the average age at which they give birth. "Over time, women get access to education, they have positions outside the family, jobs, all of those that compete with childbearing," says Ezeh.

However, Ezeh is keen to emphasize the merits of education independent from their impact on population size – it's one of the UN's 17 sustainable development goals. This gets to the heart of one modern view on population engineering – policies should be implemented for the benefit of society, and if they happen to lead to beneficial demographic changes, that's just a bonus.

"I think one of the things we don't want to do is to instrumentalize female education and make it that we want them to go to school because we want women to have fewer children… there are a lot of positives that we cannot minimize by thinking about it in terms of fertility reduction," says Ezeh.

In fact, the cascading side-effects of policies implemented for other reasons highlight a striking reality of population science: just how imprecise its predictions often are. Across the globe, the decisions made by governments over the coming decades might be, in some way, hugely influential in determining how many people there are on the planet – with the power to bump us from a future in which there are 10 billion people to one in which there are 15 billion, and vice versa.

"I think one of the things we know for sure is… [when people say that] the population of Africa is projected to be x in year y, [it] is not a destiny," says Ezeh. "If you look at the Southern Africa region… its population could be three to four times larger than what it is today by 2100, but it could also be less than 50% larger than what it is today by 2100. It's such a broad range of possibilities - whether we make the necessary investments to get to the growth rate consistent with where countries want to go. So that is the magnitude of the opportunity that exists."

An Expanding Presence

However, though the degree to which humanity will continue to expand across the planet is still to be decided, some trajectories have already been set. One is that the human population will likely continue to grow for some time, regardless of any possible efforts to decrease it.

This future is down to a phenomenon known as "demographic momentum," in which a young population with a fertility rate below replacement level will continue to grow as long as the death rate and migration levels remain the same. This is because population change isn't just about birth rates – the structure of a population also has an impact, particularly the total number of women at childbearing age. All this means that in countries where fertility rates are high, the full impact of this growth isn't felt until the women in that population reach reproductive age decades later.

One 2014 study found that even in the event of a major global tragedy such as a deadly pandemic, catastrophic world war, or draconian one-child policy implemented in every country on the planet – none of which anyone is hoping for, of course – our population will still grow up to 10 billion people by 2100. Even a disaster on such a scale that it leaves two billion people dead within a five-year period in the middle of the century would still see the population grow by 8.5 billion people by 2100. Whatever happens, the authors conclude, there are likely to be many people around until at least the next century.

With humanity set to become even more dominant in the years to come, finding a way to live together and protect the environment could be our species' greatest challenge yet.

What Does Climate Change Actually Do to Us?

- Higher-than-average temperatures and sweltering heat waves can cause fatalities.

- Longer-lasting droughts in the American West: A megadrought lasting 22 years was recorded as the worst in 1,200 years.

- Shrinking water supplies. For example, Colorado's water trickles when it reaches the Imperial Valley. The Western Water Right needs review. Reno, Nevada, needs to curtail its use of entertainment, for starters, and other large water users need to be reviewed.

- Forests are becoming insect-ridden.

- More intense forest fires. California, Oregon, and Washington desperately need to take care of their forests to reduce massive carbon fire issues and help with fatalities. 19 square miles of forest were destroyed recently.

- Stronger storms, with each degree of 7 percent Celsius, causing flooding in low areas of states, including roads and tunnels. An incredible 2017 report details massive rainfall with Category 4 and 5 storms that laid down 275 trillion pounds of rain on Houston, causing dozens of deaths.

- Records are seemingly being smashed worldwide.

- Flooding of floodplains is expected to grow to 45% by 2100.

- In Pakistan, one-third of the country is flooded with torrential rains.

- 70% of the planet is covered in water, and 93% of the Sun's warmth warms it.

- Greenhouse gases and 30% of all carbon dioxide are emitted from burning fossil fuels in woodlands in South and Central America and around the world.

- Warmer ocean waters around the world are also being affected, causing sensitive fish to move and plants, coral, and marine life to die.

- Not mentioned in the articles was the fact that most New Hampshire ponds and lakes are acidic and will not support fish. New Hampshire and parts of Maine are considered the butthole of the nation because of the airflow from industrialized states. Globally, marine life is also affected, as corals, oysters, and clams have not fully developed their shells.

Author's Notes:

20,000 young people from 10 different countries were surveyed, and the results were published in Nature Magazine in 2021. 45% said that their feelings concerning climate change varied from anxiety to powerlessness to anger, impacting their lives. This is a good sampling, and I have experienced their angst.

In a public opinion poll, the Washington Post and the Kaiser Family Foundation conducted a poll in November of 2019 (yes, it is dated but important). The poll revealed that 3 in 4 Americans heard little or nothing about the NGD. Of those who said they had heard a good deal, 6 of 10 opposed it. Among the adults, 20 percent supported it, 20 percent opposed it, and 60 percent had no opinion. The importance of these polls suggests at the onset that the public either did not trust the government or the program was too aggressive. They did not buy dumping all the fossil fuels right up front, and training the workers did not make sense.

A Christian from the South said, "To me, it is too much, too fast about the GND. If it was a 100- or 50-year goal presented, then let's work on it."

The proposal contains potential costs, and 67% of Americans said they would oppose the NGD if federal spending was in the Trillions. Biden cost $1.7 Trillion, Warren cost $2 Trillion, and Sanders cost $16.3 Trillion. I guess that says a lot to the voting public. I don't wonder why this was never put to a vote and is only a resolution in every iteration to mask rampant spending for what turned out to be largely ineffective.

Under worsening inequity, the wealthy nations, read, we are a rich nation and prime to be picked clean as you will learn. This is why we are said to, such as the US, emit more greenhouse gas emissions. I wholeheartedly disagree, and if this is the position of our New Green Deal administrators, then they need to survey China, Russia, India, and those who do not adhere to the problematic issues. I agree there are inequalities in states, which is a local and state issue until reasonable minds can devise a workable solution.

If you think the people in our nation's legislative branches are not the real driving force from which all the planning seems to come, I will reveal this group and where this movement's impetus is emanating. You will be dumbstruck! They are prepared to take all your money, and I believe you will not have much to say. A world assembly ready to take all your money for world order is not being addressed just yet. I have the facts you will learn that the US is a pawn to this organization; more on this later.

Effects Of Climate Change on Humans

Humans feel the same effects as all the other categories - weather, animals, and agriculture. But it could and has and will be a reality and might alter the fabric of society. It could lead to famine, disease, war, displacement, injury, and possibly death. For an extremely large segment of the world population, this is reality. For the fortunate, it may seem to be an inconvenience. Climate change has a hold on us that has existed for thousands of years already and, without a doubt, in my mind, is an existential threat to all human life.

Author's Notes:

For some issues, there may not be a reversal of conditions; some would wish they could go away, but being a practical person, you will be shocked to the core after you read some of the things I will present. Fact: polar ice will continue to disappear, and misuse of underground water resources, like the Ogallala Aquifer, is escalating. Massive deep mines, often stretching 1–2 miles deep and spanning miles wider to extract critical metals important to economies, will be extremely hard to control. There are many more such projects all over the globe all having an effect on the worlds stability.

I draw your attention to Archelogy Magazine, January/February 2024, Pages 38-41. Thanks to the publication and to all the contributors to the article. This is an incredibly special entry concerning something we could NOT CONTROL with the New Green Deal. This is the Earth and Earth mechanics; it rules!

This is a true example of something we could be hard-pressed to believe, but the Native tribes - Washoe, Western Shoshone, Ute, and Piute – still live (yes, they still do) in the Great Basin. This area has a closed watershed; all water from the surrounding mountains drains into lakes but does not flow into the ocean. For millennia, animals and landscape plants have supported foraging communities and have been for 2000, or as much as 3000 years, in encampments with shelter and evidence of processing food.

In the Late Holocene Dry period 3000 years ago, the Pacific Ocean became exceptionally warm, kicking off La Nina climate events in parts of the Northwest. In the 1000 years since this event, people have had to change their lives to tough it out. Chippewa scholar Gerald Vizenor commented: "The idea of survival suggests that in a bad situation, people put their heads down and try to get through on a day-by-day existence."

Thomas from the Museum of Natural History states, "Survivivance, on the other hand, is survival with an attitude. No matter how bad the situation gets, people persevere to maintain their culture so they can return.

Author-- Plagiarism and a Qualifying Statement:

I am not a scientist, just a business guy. I draw down articles, excerpts, and opinions that are meant to inform with attribution and thanks. I do not want the reader to think I am plagiarizing. While I understand the presentations, I am not thoroughly versed or qualified in the Earth Sciences fields. I know that the articles that I present are important to the point of responsible disclosure in evaluating a concept or program, and you absolutely need referential and detailed points of view to understand what is transpiring.

Human Health Is in the Equation as Well

Climate change worsens air quality, and sickness can come from wildfire smoke. Ozone smog in cities affects allergies and preexisting conditions like asthma and heart disease.

Insect-borne diseases like Malaria (in my infancy, I contracted Encephalitis, an insect-borne disease that disabled me for five years, taking me five years to recover). Malaria originated from the hot climate of the Philippines, where heat air temperatures typically can be 120-135 degrees—Lyme disease from ticks, and Zika-like warm weather.

Other issues include heat stress, stagnant water, and unsanitary conditions that seem to emerge from oppressive heat. The poor suffer when these conditions exist.

Permafrost, Our MAJOR Climate Control KILLER!

It's not extremely helpful, but when the ice melts from the permafrost layers, which it does, vast amounts of carbon dioxide are released into the atmosphere. This is unavoidable and uncontrollable. It is not a minor problem and has a major downside, and up to the year 2100, it seems to be especially important.

Permafrost covers approximately 22.8 million square kilometers (about 8.8 million square miles) in Earth's Northern Hemisphere. The downside is that the New Green Deal cannot control the melting or the release of dangerous and harmful emissions. Is this factored into their overall calculations, I think not, a fundamental omission of facts to mislead, to push their flawed agenda?

Author notes

SERIOUS IS NOT AN APPROPRIATE WORD. DISASTER IN MAKING IS!!

When I discussed with you that the NGD program was superficial and lacked an understanding of the realities of eventual impacts on many planes, this below is a prime example of what I was referring to.

AI Overview

Permafrost thaw has several serious downsides, including accelerating climate change by releasing greenhouse gases like methane and carbon dioxide, damaging infrastructure, and destabilizing ecosystems. It also poses a risk of releasing dormant viruses and bacteria.

Here's a more serious, detailed breakdown, and it's not just releasing gases; it's much more serious:

- Climate Change Acceleration:

As permafrost melts, it releases vast amounts of methane and carbon dioxide, which are potent greenhouse gases that contribute to global warming. This creates a positive feedback loop, as increased warming further accelerates permafrost thaw.

- Infrastructure Damage:

Thawing permafrost can cause considerable damage to buildings, roads, pipelines, and other infrastructure built on frozen ground. The ground can become unstable, leading to sinkholes, cracks, and collapses.

- Ecosystem Disruption:

Thawing permafrost alters natural landscapes, creating thermokarsts (areas of sunken ground and shallow ponds) and increasing the risk of landslides and coastal erosion. This can affect s warming aquatic ecosystems, degrade water quality, and harm wildlife.

- Release of Pathogens:

Permafrost contains ancient microorganisms and pathogens, some of which could be released as permafrost thaws, posing a threat to human and animal health that we have no knowledge to control.

- Economic Costs:

The damage caused by permafrost thaw can result in significant economic losses, including infrastructure repair costs and potential impacts on agriculture and other industries.

- Social Disruption:

Thawing permafrost can disrupt the lives of communities in the Arctic, particularly Indigenous communities who rely on the stability of frozen ground for housing, infrastructure, and subsistence.

Thawing of the massive permafrost- A pending problem

The permafrost is warming faster than average. This leads to thawing of the soil. This leads to the thawing of permafrost soil causing the release of carbon it has stored for thousands of years and threatening to accelerate climate change further. Also, the frozen soils has mainly yielded well-preserved megafauna remains that are releasing dangerous pathogens, such as anthrax.

It's dangerous when the release of extremely substantial amounts of carbon dioxide into the atmosphere is not helpful, but when the ice melts from the permafrost layers, which it does, vast amounts of carbon dioxide are released into the atmosphere. We cannot stop this so it's unavoidable and if the NGD folks were to draw down funds to alleviate this, they would find it uncontrollable. This is unavoidable and uncontrollable. Moving forward to their 2100 target for controls, they would find it has a major downside to being controlled and it would be VERY EXPENSIVE if not impossible.

Author's Opinion:

If you have thus far paid attention to what I am writing about green gases, while they are identified as the culprits in our atmosphere, they seem to be reduced only by the full contribution of the entire world. This simple fact proves to me that the stated final desired outcomes of the New Green Deal cannot and will not ever be realized.

I have not yet seen or heard of a comprehensive plan of action released to the public concerning the objectives of the NGD or an admission of the acknowledgment that uncontrollable actions exist. Fess up, NGD folks!

Earth's Axis Has Shifted – Sadly, It's Mostly Our Fault

According to a study published in Geophysical Research Letters in June 2023, humans have shifted Earth's rotational axis by more than 31 inches (78.7 centimeters) in less than 20 years due to groundwater extraction.

- Earth's axis has shifted due to climate change.

- Melting glaciers and overuse of groundwater account for much of the change.

- Regions like Alaska and the Himalayas have experienced the most glacial melting.

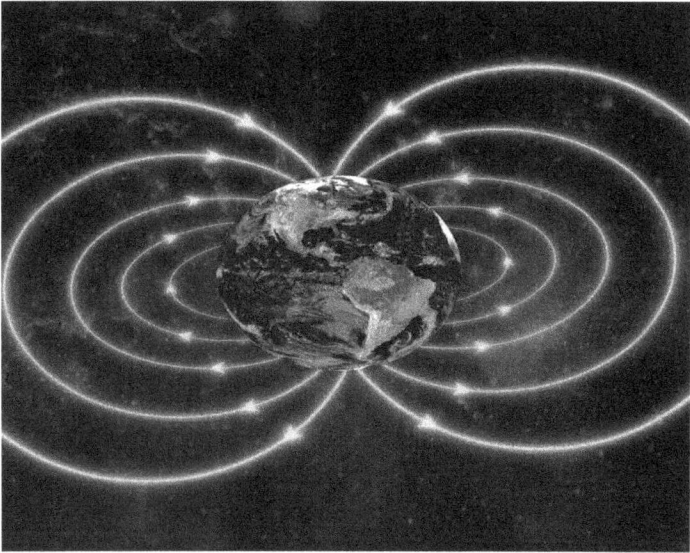

Melting Glaciers Have Shifted the Earth's Axis

In the 1990s, the Earth's axis underwent a major shift. It is normal for the Earth's axis to move by a few centimeters each year. But, in the 1990s, the direction of polar drift shifted suddenly, and the rate of the drift accelerated. The reason for this sudden change was previously unclear, but a team of scientists in Beijing recently released a paper that shows that the main driver of the change in the direction of the axial shift was glacier melt caused by global warming.

Earth as seen from a satellite.

The Earth's spin axis is the figurative line about which the Earth rotates. The poles, north and south, are situated at either end of the spin axis. By contrast, the magnetic poles - the ones you can find using a compass - are usually offset from the geographic poles, and their location shifts with the magnetic field.

"The Earth rotates around its axis somewhat like a spinning top," explained Suxia Liu in an interview with Glacier Hub. "If the weight from one area is moved to another, the spinning top will start to lean, causing the rotation axis to change." Liu co-led the paper with colleague Shanshan Deng at the Chinese Academy of Sciences.

Bernhard Steinberger, a researcher at the GFZ German Research Centers for Geosciences, explained how glaciers influence mass distribution. "The Earth always orients itself relative to the pole in a way to move masses as far as possible away from the pole," he wrote in an interview with Glacier Hub. "For example, if a glacier grows on Greenland, the Earth's orientation will change so that Greenland is further away from the pole. If a glacier melts in Greenland, it will change in the opposite direction."

Before human-caused forces, the primary drivers of polar drift were ocean currents and the movements of the molten rock deep below the Earth's surface. The research team reanalyzed existing data to determine what role terrestrial water storage (how water is dispersed above Earth's surface and in oceans and groundwater) played in the shift. They determined that the key driver of the directional change was glacier melt and the change in mass distribution that arose from it.

Most of the world's glaciers are above ground. They melt, and the water they contain moves into bodies of water. "Shifting water storage away from above-ground glaciers in one area on the Earth's surface to other results in the polar shift due to the weight change," said Liu.

An aerial view: The Greenland Ice Sheet is dotted with blue meltwater 'lakes.'

In 1995, the direction of the planet's polar drift abruptly shifted from southward to eastward. Today, scientists can connect polar drift to glacier loss using gravitational data from NASA's twin Gravity Recovery and Climate Experiment (GRACE) satellites launched in 2002. However, since the satellites had not yet been launched in the 1990s, researchers had to piece together the reasons for the sudden directional shift without these detailed gravitational records.

Besides the greenhouse gas emissions that cause glaciers to melt, other human activity is responsible for changes in Earth's hydrosphere. The team also found that groundwater pumping was, and continues to be, a factor in polar drift. Groundwater pumping for drinking water, irrigation, and manufacturing has been common since the 1960s. The United States uses 82.3 (2021) billion gallons of groundwater daily. The water removed from below ground ends up in the atmosphere as it evaporates from irrigated crops or in the ocean as runoff from irrigation systems into rivers, redistributing mass worldwide and altering the planet's rotation.

A groundwater pumping station in California's Central Valley.

Ancient plant and insect bites confirm Greenland melted in recent geologic past.

As the Earth's glaciers recede at unprecedented rates, the planet's mass is constantly redistributed. The team's findings suggest that the hydrosphere will continue to cause the Earth's axis to shift in the coming years.

Though imperceptible to humans without the use of specialized instruments, the tremendous mass of the Earth has been shifting more than ever recorded. Since 2005, the rate of polar drift has increased by about four centimeters per year. "The shift is nothing that a regular person might notice in their day-to-day life," said Steinberger. "One would have to wait millions of years to notice something." It is, however, a stark reminder of the magnitude of humans' effects on the planet.

Planting Some Tree Species May Worsen, Not Improve, NYC Air, Says New Study

Interesting Comments

J Senkyrik

It won't be a problem for the existence of the Earth itself, but it is a cataclysmic problem for life on the surface of the Earth. We will see even greater natural disasters yet to happen. No amount number of wishful-thinking comments will have any effect on this.

Phil Gman

A laughable analysis by China – a country that has never applied a doctrine to reduce Climate Change. Fact: they break every agreement purposefully. China uses climate change to reduce geopolitical rival output, thus increasing its stranglehold on the world economy, all while hanging the carrot of research grant money in front of all eager research institutes.

Climate Change is entirely unstoppable, and the speed of change is merely an extrapolation of consequential acts, resulting from population density or application of carbon-based energy development—it's simply when and how not if. The Earth is in a constant state of change at all times. As little as 500 years ago, Europe went through drastic temperature changes, entirely independent of mankind's actions.

The world has always been in a precarious state of balance, and minor changes in climate have had dramatic effects, even within the last 100 years. So rather than wasting effort trying to stop an unstoppable series of events, know these changes are going to occur and take action, anything else is simply hyperbola.

Permafrost is melting, ice sheets are breaking off, and glaciers are receding. Personally, I am more fearful of the impact on humanity that even a mini-ice age event would have than a warming event—nuclear winter included.

Sea levels are rising - take action to avert that certainty now. Potable water is one shortage today, but the change in air-able land with a reliable water source is more important. We as a species are facing a forceable world famine due to reduced output – another certainty. It only took the loss of one crop planting in Ukraine over a period of the last 2 months to ensure a world food shortage for some and starvation for many others. Yet, real action today again can avert a complete catastrophe.

Chance

Scientists have found that a shift in Earth's poles 42,000 years ago may have drastically altered the planet's climate—and they're naming the period after the author Douglas Adams. About 42,000 years ago, Earth was beset with oddness. Its magnetic field collapsed. Yes, it works in the Adams Event Las Champ cycle. There are so many variables from so long ago. It takes a broad-spectrum analysis of everything that happened in conjunction. However, the polar shift is left out of the mainstream, requiring a deeper discussion of preparation as it is clearly changing.

Sue

Isn't this what happened millions of years ago? The planet went through ice ages and hot climates, during which most life was lost.

How can you stop the sun from shining?

Just because it's shining over the North Pole doesn't mean it can be protected, and everybody knows that ice will melt over time.

Isn't there some things, like day and night coming and going, that can't be prevented?

Some people are producing ideas, but many people have different ideas. It has a lot to do with the heating, but our planet's axis shifts and our orbit around our sun may be wobbling, which is likely.

When you combine all the factors and the increased heating of our planet, we should look to Australia, where many inhabitants live in either caves or underground homes. Many miners live in dugouts, and the temperature maintains a constant cooler temperature.

Rachel Wiggins

I have been deeply intrigued by this subject since I read the book 'Pole Shift' by John White, "Worlds in Collision" by Immanuel Velikovsky, and some studies by Charles Hapgood back in the 1990s. the glacier's melting has something to do with it as the article describes it becoming offset. Another good example is your washer; when the clothes are too much on one side, and the washtub gets off balance, it starts to wobble, and an alarm will sound. The alignment of planets and our sun and moon play a big part in this shift. Some movies explain it well. Is it possible that they may be right after all?

Zen

Are the axes shift small? Then, explain to me how people are now waking up to the sun brightening their rooms from the north. For those who don't know, the sun has been appearing from the North in Dallas-Fort Worth.

The sun is supposed to be directly over the Tropic of Cancer on the Summer Solstice on June 21st. That is the farthest north position of the sun directly overhead. The Tropic of Cancer is near central Mexico. The sun has NEVER been any farther north until now.

We should never see the sun rising and setting to the north of the Texas border… but it is. You do realize that extended sunlight can also be an indication of a big axis shift. The changes in big shifts can be seen in seasons and in some parts of the globe.

A GIF showed an abrupt change of the axis because the satellite weather map had changed; now, I can't find it. They deleted it.

Christina Sosseh

The Times of India reported that the 2004 Asian tsunami clipped Earth's Day by 6.8 microseconds and shifted its axis by about 3 inches. The Chilean earthquake in 2010 also shortened the day by

1.26 microseconds and changed the axis by 3 inches. The shifting of islands was observed in the 2004 tsunami, too.

Mac

Has anyone discussed the theory that the axis shift causes changing patterns and more extreme weather conditions, as opposed to humanity or cows/farming causing "global warming?" This planet has gone through many violent and extreme changes throughout its life. It will continue to change long after we're gone. We are not the cause. We're like a rash on a dog's back—irritating but not deadly.

Why Are Glaciers and Sea Ice Melting?

What Is the Difference Between Sea Ice and Glaciers?

Sea ice forms and melts strictly in the ocean, whereas glaciers are formed on land. Icebergs are chunks of glacial ice that break off glaciers and fall into the sea.

When glaciers melt, because that water is stored on land, the runoff significantly increases the ocean's water, contributing to global sea level rise.

is often compared to ice cubes in a glass of water: when it melts, it does not directly change the water level in the glass. Instead, depleting Arctic sea ice triggers a host of other devastating consequences, from depleting available ice on which walruses can haul out or polar bears hunt to changing weather systems around the world by altering the pattern of the jet stream.

Why Are the Glaciers Melting?

Since the early 1900s, many glaciers worldwide have been rapidly melting. Human activities are at the root of this phenomenon. Specifically, since the industrial revolution, carbon dioxide and other greenhouse gas emissions have raised temperatures even higher in the poles. As a result, glaciers are rapidly melting, calving off into the sea, and retreating on land.

Even if we significantly curb emissions in the coming decades, more than a third of the world's remaining glaciers might melt before the year 2100. Regarding sea ice, 95% of the oldest and thickest ice in the Arctic is already gone.

Scientists project that if emissions continue to rise unchecked, the Arctic could be ice-free in the summer as soon as 2040, as ocean and air temperatures continue to rise rapidly.

What Can I Do to Help?

Strong action on climate change means preparing communities for the impacts that are happening now. But it also means looking to the future, focused on reducing the heat-trapping gases in our atmosphere that will bring damaging consequences as our planet warms.

The good news is that individuals can play a big part.

Reach out to your local elected officials to find out if your city has a disaster response plan for right now. Keeping communities safe starts with having a strong plan in place. Communities start with a strong plan that leverages some of the best but underutilized tools we have to protect our communities: nature.

What Are the Effects of Melting Glaciers on Sea Level Rise?

Melting glaciers contribute to rising sea levels, increasing coastal erosion, and elevating storm surges as warming air and ocean temperatures create more frequent and intense coastal storms like hurricanes and typhoons. Specifically, the Greenland and Antarctic ice sheets are the largest contributors to global sea level rise. Right now, the Greenland ice sheet is disappearing four times faster than in 2003 and already contributes 20% of the current sea level rise.

How much and how quickly these Greenland and Antarctic ice sheets melt in the future will largely determine how much ocean levels rise. If emissions continue to rise, the current melting rate on the Greenland ice sheet is expected to double by the end of the century. Alarmingly, if all the ice in Greenland melted, global sea levels would rise by 20 feet.

How Do Melting Sea Ice and Glaciers Affect Weather Patterns?

Today, the Arctic is warming twice as fast as anywhere on Earth, and the sea ice there is declining by more than 10% every 10 years. As this ice melts, darker patches of the ocean start to emerge, eliminating the effect that previously cooled the poles, creating warmer air temperatures and, in turn, disrupting normal patterns of ocean circulation. Research shows the polar vortex is appearing outside of the Arctic more frequently because of changes to the jet stream caused by a combination of warming air and ocean temperatures in the Arctic and the tropics.

The glacial melt we are witnessing today in the Antarctic and Greenland is changing the circulation of the Atlantic Ocean and has been linked to the collapse of fisheries in the Gulf of Maine and more destructive storms and hurricanes around the planet.

What Are the Effects of Melting Glaciers and Sea Ice Loss on Humans and Wildlife?

What happens in these places has consequences across the entire globe. As sea ice and glaciers melt and oceans warm, ocean currents will continue to disrupt weather patterns worldwide. Industries that thrive on vibrant fisheries will be affected as warmer waters change where and when fish spawn. Coastal communities will continue to face billion-dollar disaster recovery bills as flooding becomes more frequent, and storms intensify. People are not the only ones impacted. In the Arctic, as sea ice melts, wildlife like walrus are losing their home, and polar bears are spending more time on land, causing higher rates of conflict between people and bears.

Meltwater gushes from an ice cap on the island of Nordaustlandet in Norway's Svalbard archipelago.

The Big Thaw

How much and quickly will Earth's glaciers melt as the climate warms?

Many thanks to National Geographic for a great teaching tool and to Daniel Glick, our guide.

"If we don't have it, we don't need it," pronounces Daniel Fagre as we throw on our backpacks. We're armed with crampons, ice axes, rope, GPS receivers, and bear spray to ward off grizzlies, and we're trudging toward Sperry Glacier in Glacier National Park, Montana. I fall in step with Fagre and two other research scientists from the U.S. Geological Survey Global Change Research Program. They're doing what they've been doing for over a decade: measuring how the park's storied glaciers are melting.

So far, the results have been positively chilling. When President Taft created Glacier National Park in 1910, it was home.

"Things that normally happen in geologic time are happening during the span of a human lifetime," says Fagre. "It's like watching the Statue of Liberty melt."

Author Disclosure:

I am not trained in this field and prefer giving my reader opinion to an estimated 150 glaciers. Since then, the number has decreased to fewer than 30, and most of those remaining have shrunk in area by two-thirds. Fagre predicts that within 30 years, most, if not all, of the park's namesake glaciers will disappear. This book presents unaltered facts rather than offering untrained and possibly incorrect guidance, as I aim to address what I feel is deleterious to our common lifestyles by driven individuals. I fear that they are less conversant than I am about crucial matters that are critical to our government, spending, and our future.

Climate 101: Glaciers

Scientists who assess the planet's health see indisputable evidence that Earth has been getting warmer, in some cases rapidly. Most believe that human activity, in particular the burning of fossil fuels and the resulting buildup of greenhouse gases in the atmosphere, has influenced this warming trend. In the past decade, scientists have documented record-high average annual surface temperatures and have been observing other signs of change all over the planet: in the distribution of ice and in the salinity, levels, and temperatures of the oceans.

"This glacier used to be closer," Fagre declares as we crest a steep section, his glasses fogged from exertion. He's only half joking. A trailside sign notes that since 1901, Sperry Glacier has shrunk from more than 800 acres (320 hectares) to 300 acres (120 hectares). "That's out of date," Fagre says, stopping to catch his breath. "It's now less than 250 acres (100 hectares)."

Everywhere on Earth, ice is changing. The famed snows of Kilimanjaro have melted more than 80 percent since 1912. Glaciers in the Garhwal Himalaya in India are retreating so fast that researchers believe that most central and eastern Himalayan glaciers could virtually disappear by 2035. Arctic sea ice has thinned significantly over the past half-century, and its extent has declined by about 10 percent in the past 30 years. NASA's repeated laser altimeter readings show the edges of Greenland's ice sheet shrinking. Spring freshwater ice breakup in the Northern Hemisphere now occurs nine days earlier than it did 150 years ago, and autumn freeze-up ten days later. Thawing permafrost has caused the ground to subside more than 15 feet (4.6 meters) in parts of Alaska. From the Arctic to Peru, from Switzerland to the equatorial glaciers of Man Jaya in Indonesia, massive ice fields, monstrous glaciers, and sea ice are disappearing fast.

When temperatures rise and ice melts, more water flows to the seas from glaciers and ice caps, and ocean water warms and expands in volume. This combination of effects has played a key role in raising the average global sea level between four and eight inches (10 and 20 centimeters) in the past hundred years, according to the Intergovernmental Panel on Climate Change (IPCC).

Scientists point out that sea levels have risen and fallen substantially over Earth's 4.6-billion-year history. But the recent rate of global sea level rise has departed from the average rate of the past two to three thousand years and is rising more rapidly - about one-tenth of an inch a year. A continuation or acceleration of that trend has the potential to cause striking changes in the world's coastlines.

Driving around Louisiana's Gulf Coast, Windell Curole could guess the future, and it seemed pretty wet. In southern Louisiana, coasts are literally sinking by about three feet (a meter) a century, a process called subsidence. A sinking coastline and a rising ocean combine to yield powerful effects. It's like taking the global sea-level-rise problem and moving it along at fast-forward.

The seventh-generation Cajun and manager of the South Lafourche Levee District navigates his truck down an unpaved mound of dirt that separates civilization from inundation, dry land from a swampy horizon. With his French-tinged lilt, Curole points to places where these bayous, swamps, and fishing villages portend a warmer world: his high school girlfriend's house partly submerged, a cemetery with water lapping against the white tombs, his grandfather's former hunting camp now afloat in a stand of skeleton oak snags. "We live in a place of almost land, almost water," says the 52-year-old Curole.

Rising sea level, sinking land, eroding coasts, and temperamental storms are an unpleasant fact for Curole. Even relatively small storm surges in the past two decades have overwhelmed the system of dikes, levees, and pump stations that he manages, upgraded in the 1990s to forestall the Gulf of Mexico's relentless creep. "I've probably ordered more evacuations than any other person in the country," Curole says.

The current trend is consequential not only in coastal Louisiana but around the world. Never before have so many humans lived so close to the coasts: More than a hundred million people worldwide live within three feet (a meter) of mean sea level. Vulnerable to sea-level rise, Tuvalu, a small country in the South Pacific, has already begun formulating evacuation plans. Megacities where human populations have concentrated near coastal plains or river deltas - Shanghai, Bangkok, Jakarta, Tokyo, and New York - are at risk. The projected economic and humanitarian impacts on low-lying, densely populated, and desperately poor countries like Bangladesh are potentially catastrophic. The scenarios are disturbing even in wealthy countries like the Netherlands, with nearly half its landmass already at or below sea level.

Rising sea level produces a cascade of effects. Bruce Douglas, a coastal researcher at Florida International University, calculates that every inch (2.5 centimeters) of sea-level rise could result in eight feet (2.4 meters) of horizontal retreat of sandy beach shorelines due to erosion. Furthermore, when salt water intrudes into freshwater aquifers, it threatens sources of drinking water and makes raising crops problematic. In the Nile Delta, where many of Egypt's crops are cultivated, widespread erosion and saltwater intrusion would be disastrous since the country contains little other arable land.

In some places, marvels of human engineering worsen the effects of rising seas in a warming world. The system of channels and levees along the Mississippi effectively stopped the millennia-old natural process of rebuilding the river delta with rich sediment deposits. In the 1930s, oil and gas companies began to dredge shipping and exploration canals, tearing up the marshland buffers that helped dissipate tidal surges. Energy drilling removed vast quantities of subsurface liquid, which studies suggest increased the rate at which the land is sinking. Now Louisiana is losing approximately 25 square miles (65 square kilometers) of wetlands every year, and the state is lobbying for federal money to help replace the upstream sediments that are the delta's lifeblood.

Author note: My question and possible solution to the above issue is to have the people causing the problem take care of it by fixing the damage they are doing as a cost of doing business. The state is probably getting fees from the companies. Why should the state ask for federal funds when they are probably getting funds from the developers? Folks we are all paying for the state allowing this to happen without tax dollars and getting no benefit from this likely scheme.

Local projects like that might not do much good in the extraordinarily long run, though, depending on the course of change elsewhere on the planet. Part of Antarctica's Larsen Ice Shelf broke apart in early 2002. Although floating ice does not change sea level when it melts (any more than a glass of water will overflow when the ice cubes in it melt), scientists became concerned that the collapse could foreshadow the breakup of other ice shelves in Antarctica and allow increased glacial discharge into the sea from ice sheets on the continent. If the West Antarctic ice sheet were to break up, which scientists consider very unlikely this century, it alone contains enough ice to raise sea level by nearly 20 feet (6 meters).

Even without such a major event, the IPCC projected in its 2001 report that the sea level would rise between 4 and 35 inches (10 and 89 centimeters) by the end of the century. The high end of that projection - nearly three feet (a meter) - would be "an unmitigated disaster," according to Douglas.

Down on the bayou, all of those predictions make Windell Curole shudder. "We're the guinea pigs," he says, surveying his aqueous world from the relatively lofty vantage point of a 12-foot-high (3.7-meter) earthen berm. "I don't think anybody down here looks at the sea-level-rise problem and puts their heads in the sand." That's because soon there may not be much sand left.

Rising sea levels are not the only change Earth's oceans are experiencing. The ten-year-long World Ocean Circulation Experiment, launched in 1990, has helped researchers better understand what is now called the ocean conveyor belt.

Oceans, in effect, mimic some functions of the human circulatory system. Just as arteries carry oxygenated blood from the heart to the extremities and veins return blood to be replenished with oxygen, oceans provide life-sustaining circulation to the planet. Propelled mainly by prevailing winds and differences in water density, which changes with the temperature and salinity of the seawater, ocean currents are critical in cooling, warming, and watering the planet's terrestrial surfaces - and in transferring heat from the Equator to the Poles.

The Larsen C Ice Shelf collapse in 2017, which was the size of Los Angeles, may be a sign of a continent-wide collapse that could threaten coastal cities worldwide. The shelf's deterioration has been ongoing for over 20 years, and some factors that may have contributed to its collapse include:

Warm Summers

A series of warm summers on the Antarctic Peninsula, especially in 2002, caused significant surface melting, creating melt ponds. These ponds acted like wedges, deepening crevasses and eventually causing the shelf to splinter.

Warm Ocean Temperatures

Warm ocean temperatures in the Weddell Sea during the same period may have caused the ice shelf's underside to thin and melt.

Atmospheric Rivers

In March 2022, temperatures in Antarctica rose to 38 degrees Celsius above normal, likely due to an atmospheric river, a long plume of moisture that transports warm air and water vapor from the tropics. The warmer temperatures and rainfall associated with atmospheric rivers can cause surface melt that weakens ice from the top down.

Foehn Winds

These warm, dry winds are formed when air is forced over steep mountains, such as those on the Antarctic Peninsula. When the wind descends on the other side of the mountain, it brings in warmer air that can cause melting, even in the cooler seasons of spring and autumn.

The collapse of the Larsen Ice Shelf could lead to further loss of ice, which could accelerate inland glaciers and contribute to sea-level rise. The shelf holds back the equivalent of about 1 cm of global sea-level rise, which is similar to the rate at which sea levels are currently rising due to global warming.

The engine running the conveyor belt is the density-driven thermohaline circulation for heat and 'halite' for salt). In surface currents like the Gulf Stream, warm, salty water flows from the tropical Atlantic north toward the Pole. This saline water loses heat to the air as it is carried to the far reaches of the North Atlantic. The coldness and high salinity make the water denser and sink deep into the ocean. Surface water moves in to replace it. The deep, cold-water flows into the South Atlantic, Indian, and Pacific Oceans, eventually mixing again with warm water and rising back to the surface.

Depending on how drastic they are, water temperature and salinity changes might have considerable effects on the ocean conveyor belt. According to scientists at the National Oceanic and Atmospheric Administration (NOAA), ocean temperatures are rising in all ocean basins and at much deeper depths than previously thought. Arguably, the largest oceanic change ever measured in the era of modern instruments is the declining salinity of the subpolar seas bordering the North Atlantic.

Robert Gagosian, president and director of the Woods Hole Oceanographic Institution, believes that oceans hold the key to potential dramatic shifts in the Earth's climate. He warns that too much change in ocean temperature and salinity could disrupt the North Atlantic thermohaline circulation enough to slow down or possibly halt the conveyor belt - causing drastic climate changes in time spans as short as a decade.

The future breakdown of the thermohaline circulation remains a disturbing, if remote, possibility. But the link between changing atmospheric chemistry and changing oceans is indisputable, says Nicholas Bates, a principal investigator for the Bermuda Atlantic Time-series Study station. The station monitors the temperature, chemical composition, and salinity of deep-ocean water in the Sargasso Sea southeast of the Bermuda Triangle.

Oceans are important sinks, or absorption centers, for carbon dioxide and are about a third of human-generated CO_2. Data from the Bermuda monitoring programs show that CO_2 levels at the ocean surface are rising at about the same rate as atmospheric CO_2. But it is at deeper levels where Bates has observed an even greater change. In the waters between 820 and 1,476 feet (250 and 450 meters) deep, CO_2 levels rise at nearly twice the rate in the surface waters. "It's not a belief system; it's an observable scientific fact," Bates says. "And it shouldn't be doing that unless something fundamental has changed in this part of the ocean."

While scientists like Bates monitor changes in the oceans, others evaluate CO_2 levels in the atmosphere. In Vestmannaeyjar, Iceland, a lighthouse attendant opens a large silver suitcase that looks like something out of a James Bond movie, telescopes out an attached 15-foot (4.5-meter) rod, and flips a switch, activating a computer that controls several motors, valves, and stopcocks—two two-and-a-half liters (about 26 quarts) flasks filled with ambient air in the suitcase. In North Africa, an Algerian monk at Assekrem does the same. Around the world, collectors like these monitor the cocoon of gases that compose our atmosphere and permit life as we know it to persist.

When the weekly collection is done, all the flasks are sent to Boulder, Colorado. There, Pieter Tans, a Dutch-born atmospheric scientist with NOAA's Climate Monitoring and Diagnostics Laboratory, oversees several sensitive instruments that evaluate the air in the flasks for its chemical composition. In this way, Tans helps assess the state of the world's atmosphere.

By all accounts, it has changed significantly in the past 150 years.

Walking through the various labs filled with cylinders of standardized gas mixtures, absolute manometers, and gas chromatographs, Tans offers a brief history of atmospheric monitoring. In the late 1950s, a researcher named Charles Keeling began measuring CO_2 in the atmosphere above Hawaii's 13,679-foot (4,169-meter) Mauna Loa. The first thing that caught Keeling's eye was how CO_2 levels rose and fell seasonally. That made sense since, during spring and summer, plants take in CO_2 during photosynthesis and produce oxygen in the atmosphere. In the fall and winter, when plants decay, they release greater quantities of CO_2 through respiration and decay. Keeling's vacillating seasonal curve became famous for representing the Earth's "breathing."

Something else about the way the Earth was breathing attracted Keeling's attention. He watched as the CO_2 level not only fluctuated seasonally but also rose year after year. The carbon dioxide level has climbed from about 315 parts per million (ppm) from Keeling's first readings in 1958 to more than 375 ppm today. A primary source for this rise is indisputable: humans' immense burning of carbon-laden fossil fuels for their factories, homes, and cars.

Tans shows me a graph depicting levels of three key greenhouse gases - CO_2, methane, and nitrous oxide - from 1000 to the present. The three gases together help keep Earth, which would otherwise be an inhospitably cold orbiting rock, temperate by orchestrating an intricate dance between the radiation of heat from Earth back to space (cooling the planet) and the absorption of radiation in the atmosphere (trapping it near the surface and thus warming the planet).

Tans and most other scientists believe greenhouse gases are at the root of our changing climate. "These gases are a climate-change driver," says Tans, poking his graph definitively with his index finger. The three lines on the graph follow almost identical patterns: flat until the mid-1800s, then all three move upward in a trend that turns even more sharply after 1950. "This is what we did," says Tans, pointing to the parallel spikes. "We have very significantly changed the atmospheric concentration of these gases. We know their radiative properties. It is inconceivable to me that the increase would not significantly affect climate."

Exactly how large that effect might be on the planet's health and respiratory system will continue to be the subject of great scientific and political debate, especially if the lines on the graph continue their upward trajectory.

Eugene Brower, an Inupiat and president of the Barrow Whaling Captains' Association, doesn't need fancy parts-per-million measurements of CO_2 concentrations or long-term sea-level gauges to tell him that his world is changing.

"It's happening as we speak," the 56-year-old Brower says as we drive around his home in Barrow, Alaska - the United States' northernmost city - on a late August day. In his fire chief's truck, Brower takes me to his family's traditional ice cellars, painstakingly dug into the permafrost, and points out how his stores of muktuk - whale skin and blubber recently began spoiling in the fall because melting water drips down to his food stores. Our next stop is the old Bureau of Indian Affairs school building. The once impenetrable permafrost that kept the foundation solid has bucked and heaved so much that walking through the school is almost like walking down the halls of an amusement park funhouse. We head to the eroding beach and gaze out over the open water. "Normally, by now, the ice would be coming in," Brower says, scrunching up his eyes and scanning the blue horizon.

We continue our tour. Barrow looks like a coastal community under siege. The ramshackle conglomeration of weather-beaten houses along the seaside gravel road stands protected from fall storm surges by miles-long berms of gravel and mud that block views of migrating gray whales. Yellow bulldozers and graders patrol the coast like sentries.

The Inupiat language has words that describe many kinds of ice. *Piqaluyak* is salt-free multiyear sea ice. *Ivuniq* is a pressure ridge. *Sarri* is the word for pack ice, *tuvaqtaq* is bottom-fast ice, and shore-fast ice is *tuvaq*. For Brower, these words are the currency of hunters who must know and follow ice patterns to track bearded seals, walruses, and bowhead whales.

There are no words, though, to describe how much and how fast the ice is changing. Researchers long ago predicted that the most visible impacts from a globally warmer world would occur first at high latitudes: rising air and sea temperatures, earlier snowmelt, later ice freeze-up, reductions in sea ice, thawing permafrost, more erosion, and increases in storm intensity. Now, all those impacts have been documented in Alaska. "The changes observed here provide an early warning system for the rest of the planet," says Amanda Lynch, an Australian researcher who is the principal investigator on a project that works with Barrow's residents to help them incorporate scientific data into management decisions for the city's threatened infrastructure.

Before leaving the Arctic, I drive to Point Barrow alone. At the tip of Alaska, roughshod hunting shacks dot the spit of land that marks the dividing line between the Chukchi and Beaufort Seas. Next to one shack, someone has planted three eight-foot (2.4-meter) sticks of white driftwood in the sand, then crisscrossed their tops with whale baleen, a horny substance that whales of the same name use to filter life-sustaining plankton out of seawater. The baleen, curiously, looks like palm fronds.

So there stand three makeshift palm trees on the North Slope of Alaska. Perhaps they are no more than an elaborate Inupiat joke, but these Arctic palms seem an enigmatic metaphor for the Earth's future.

Popular Mechanics

New research says human-caused climate change has accelerated the rate at which Earth's rotational axis changes.

Earth has two kinds of poles. The north and south magnetic poles affect navigation, drift, and back-and-forth over time. Earth's other kind of pole is the axis around which the planet physically spins. This axis has also slightly shifted over time, but scientists haven't figured out why exactly.

The Institute of Geographic Sciences and Natural Resources Research in China and the Technical University of Denmark pulled satellite data from NASA's Gravity Recovery and Climate Experiment (GRACE) spacecraft. They studied the "true polar wander" phenomenon during a specific period in the 1990s. Scientists found that moving water masses have pushed Earth's axis eastward beyond what existing climate models predicted. On the surface, this makes sense: Think about holding a basin of water and then moving it back and forth - the sloshing causes the water's weight to move all around. This is what's happening on a macro level.

But the surprising part is that humans made it happen. The scientists took the climate model that fits data from the 2000s and 2010s and back-calculated it through to the '90s to see if the results matched up. The only way the data matches is if there's some kind of "X-factor" happening.

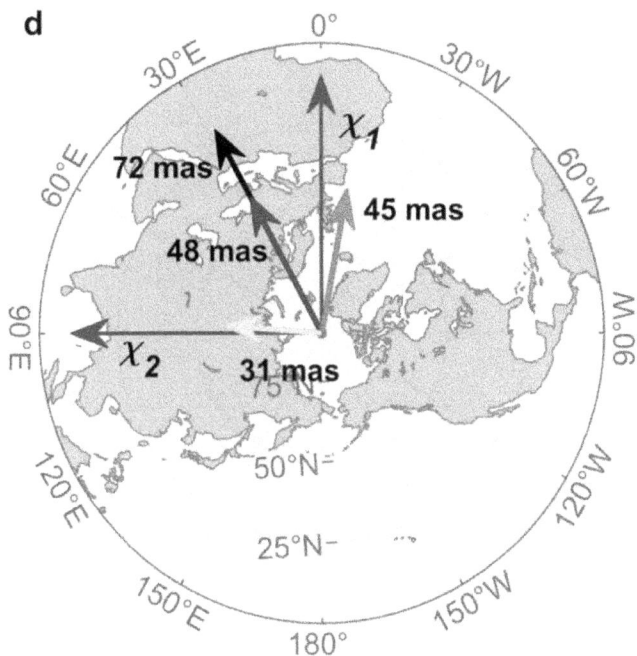

Earth's polar shift from 2002 to 2020.

That X-factor is called terrestrial water shortage (TWS), which includes changes in Earth's water levels from glacial melting and other anthropogenic activities, like overuse of the groundwater supply from aquifers. The study, which appears in *Geophysical Research Letters*, says TWS caused Earth to shift a little to the east in the 1990s - basically, the planet's full basin sloshed around and created a new center of gravity.

Now, some amount of the TWS change has naturally occurred without humans. But the scientists found a strong correlation in areas where glacial melt and groundwater depletion has happened most: "Most areas, such as Alaska, Greenland, the Southern Andes, Antarctica, the Caucasus, and the Middle East, have recently exhibited significant glacier mass changes," the scientists explain in their paper.

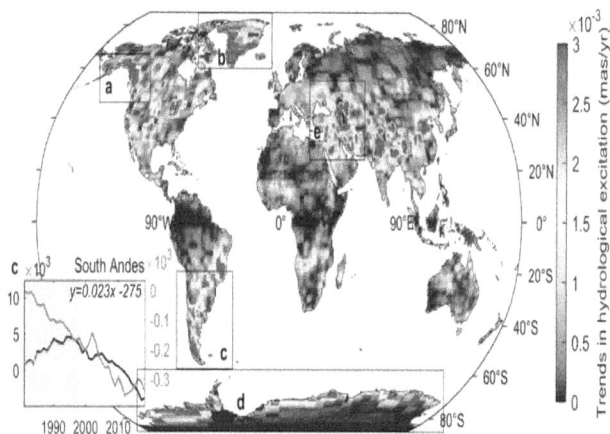

A map shows the 'hot spots' for glacial melting.

These regions, and glacial and groundwater questions more broadly, don't explain the entire disparity that scientists found between the expected values and the real values. Global water movement and storage vary by factors like regional climate patterns, they say. But overall, the mathematical analysis supports climate change factors contributing *most* to the change.

So, where does this leave us? Well, the polar shift is small in the grand scheme of things, and humans won't be able to notice it. But to make this change at such an abbreviated time indicates our near future might have bigger changes in store - ones that could make our heads spin.

The Green New Deal Is Built on Fear, Not Facts

The Green New Deal proposes to address environmental issues that are less problematic than many think.

Image credit: Pixabay. Thank you for the use of this dramatic image.

Author's Notes:

In an article I found a polling for the New Green Deal at 43.7 percent and "through the roof" The Yale Program on Climate Change and Communication provided impetus and foundational proof for this very book doubts the veracity of the claim that was made by NGD spinners   In reality it was reported that *97% of the survey respondents knew "a little or nothing at all" about it (NGD).*

Author's Note:

I am addressing the next two paragraphs. I and many others I know do not support the New Green Deal to end fossil fuel use. Why do we need to conclusively foster industries to deliver working substitute fuels? Electricity batteries have too many issues! I do, however, agree that jobs should be fostered to support any initiative. Industries in the past, if they saw a financial future, would invest in plants and assist without government interference and without having to tax the citizen base. All these GND programs seem to need BIG CASH.

As for creating jobs on a net basis, throwing bucks at the issue never seems to be a problem with the government. Let industry manage it; they are more efficient and view it as a profit opportunity, unlike feeding a bloated pig.

***

Would you support or oppose a Green New Deal to end fossil fuel use in the United States and have the government create clean energy jobs? The plan would be paid for by raising taxes, including a tax on carbon emissions?

That the question presupposes outcomes that have never happened before and are likely impossible (a massive retooling of energy provision in a decade and government creation of jobs on a net basis) is never deemed to be a problem as a proposal by the government. This proposal would never work.

The contention that climate change has resulted in stronger and/or more intense hurricanes is far from demonstrated. While one recent study purports to correlate climate change with a more rapid increase in hurricane severity since 1982 (a relatively short period given the length of the hurricane cycle and considering the National Oceanic and Atmospheric Administration, commonly known as NOAA, found the findings to be "suggestive but not definitive"). NOAA released a report, in summary, neither our model projections for the 21st century nor our analyses of trends in Atlantic hurricane and tropical storm activity support the notion that greenhouse gas-induced warming leads to large increases in either tropical storm or overall hurricane numbers in the Atlantic. Therefore, we conclude that it is premature to conclude with high confidence that human activity - and particularly greenhouse warming - has already caused a detectable change in Atlantic hurricane activity.

Of course, the duration of the power loss in Puerto Rico has more to do with the poor infrastructure than with anything else.

The notion that California wildfires are the result of climate change has largely been rejected, and globally, studies have shown that burn areas have been decreasing over the last couple of decades.

Author note 2025 addition: While this manuscript was written in 2024, a notable event concerning the California, Los Angeles, wildfire of 2025 that literally destroyed huge areas, approximately 42 square miles, and tens of thousands of destroyed civilian, public services facilities, schools, hospitals, and other infrastructure of that area was negligence, not enough water through what seems to be by design, throwing millions of gallons of water into the ocean, no water in hydrants, poor management from the Governor's administration, woke practices in Los Angeles, no fire suppression practices in the foothills, and a myriad of other issues including a person who was targeted as staring the original fire. This was not a climate control fire; it was poor administration at all levels.

While it has been argued that the 2007-2011 drought that struck Syria has exacerbated the conflict, few, if any, characterize it as a direct cause (conflict began with protests demanding democratic reforms, the release of political prisoners, and an end to corruption; religious animosity might have also played a material part). The reference to the Zika virus presupposes a northern migration of mosquitos that has not occurred and infection by a virus that has not been deemed a global threat since 2017.

The percentage of undernourished persons plummeted from 14.5 percent globally in 2005 to 10.6 percent in 2015.

As for the increase in child labor and starvation due to climate change, no such evidence exists - none. Quite to the contrary, child labor has been steadily declining, particularly in India and China, due to their greater embrace of market reforms. As for starvation, it appears that Mr. McKibben has misrepresented data from the United Nations Food and Agriculture Organization (UNFAO), which reports on 'undernourished people.' That is not the same thing.

To put things in perspective, the percentage of undernourished people globally plummeted from 14.5 percent in 2005 to 10.6

percent in 2015. Their most recent estimates put the figure at 10.9 percent in 2017. This small uptick is not attributed solely to climate change but also to multiple other causes. Conflicts around the globe, including not just Syria but Afghanistan, Yemen, Somalia, and a host of others not attributable to climate change, have been a major factor, as has the socialist disaster that is Venezuela.

Of course, if the UN's IPCC climate sensitivity figures are correct, the most one could expect from a full implementation of the Green New Deal is reduced future warming of 0.137 degrees Celsius by 2100.

Unicorns and Rainbows -Who Did the Systems Analysis?

After being shown the horror show of doing nothing, we are presented with the unicorns and rainbows of the Green New Deal.

"What the future looks like is electric cars," we are told, "[that] outperform combustion vehicles on every metric." Apparently, he has never heard of the range constraints, particularly in colder weather, of electric vehicles, or the higher average costs of repair. *Given that 3 percent of those surveyed had heard 'a lot' about the Green New Deal, it seems the most effective plan for dealing with its current popularity is to simply present the facts.*

We are told that "they cost less to own and operate," but there is a caveat that was left unmentioned. According to a University of Michigan study, it costs, on average, about 43.4 percent as much to operate an electric car than one with an internal combustion engine, as electricity is so much cheaper in relative terms.

But the likelihood of that remaining the case once the Green New Deal eliminates electricity generation from fossil fuels (currently 63 percent) and nuclear sources (currently 20 percent) rapidly approaches zero.

"What the future looks like is mass transit," we were told two days after the California governor canceled the two-billion-dollar boondoggle that was the high-speed rail project. You can't make this stuff up.

People would have you believe that adopting these things is "easy to do" and "costs pennies on the dollar compared to the future. I find it hard to accept this assessment of the future at anywhere near face value.

Given that a mere 3 percent of those surveyed had heard 'a lot' about the Green New Deal, it seems the most effective plan for dealing with its current popularity is to simply present the facts.

Pros and Cons of the New Green Deal by Real People

- Steven: All I see are a set of thoughts with not fully laid out plans.
- Ken: Without major accelerated plans for nuclear power, it's a pipe dream to replace fossil fuels with wind and solar - it would be an environmental disaster. The true goal of the NGD is political control.
- Christopher: Absolutely ZERO positive effect – it's about control. UN Agenda 21/30: The Beginning Steps of the World Economic 'Great Reset.'
- Frank: NGD passed into law—when? Develop a smart grid to manage the US. Windmills, bird graveyards… we suddenly do not care for eagles, owls, and whales. It's green, baby—green.
- Chuck: It would reduce emissions but at a huge expense to American prosperity and financial well-being. It's not feasible.
- Kenney: The NGD plan is stupid! Too costly.

Even Sponsors of the Green New Deal Can't Bring Themselves to Vote for It

It's not exactly heroic to be unwilling to stand up and vote for what you claim to believe in.

Image Credit: Pixabay - Skeeze

Author's Notes:

I find this amusing at best. I, as a former politician, have seen this 'Hot Potato' sluff off when no party actually wants to take possession of a controversial issue but wants to express themselves into a pseudo position, not to be outdone. It's comical at best, and remember, folks, you pay for them sucking up bucks and fresh air in Washington.

Senate Majority Leader Mitch McConnell (R-KY) scheduled a floor vote on the Green New Deal resolution for the week of March 24. Democrats were caught off guard when Senator McConnell said in February that he wanted to have a vote on the resolution and have been trying to figure out what to do ever since. It appears that they have now settled on a strategy: They, or nearly all of them, will vote 'present.' This includes the sponsor of Senate Resolution 59, Sen. Ed Markey (D-MA), and its eleven co-sponsors. Six of the co-sponsors are running for president.

Editor's Note:

The vote was held on Tuesday. Three Democrats voted against the Green New Deal, while the rest voted present.

Sen. Markey called the vote "nothing but an attempt to sabotage the movement we are building" in a tweet immediately after McConnell said he wanted to vote. A March 22 *Washington Post* story by Dino Grandoni quotes Markey's reasoning at length:

Democrats will not allow Leader McConnell and Republicans to make a mockery of the debate in the Senate on climate change. This vote is a sham and little more than a political ploy to protect vulnerable Republicans from having to defend their climate science denial.

As Grandon notes, it is not unprecedented for senators to be unwilling to vote for legislation that they are sponsoring. But it is odd. And it's not exactly heroic to be unwilling to stand up and vote for what you claim to believe in. What's the worst thing that could happen? Being defeated for re-election? Nor is it clear why allowing an open floor debate in the "world's greatest deliberative body" on the most ambitious energy-rationing legislation ever introduced is sabotage, a mockery, or a sham. On the other hand, having a vote is indeed a political ploy. Another undoubted political ploy was introducing the Green New Deal. It should be fun watching to find out whose political ploy turns out to be more effective.

It's Not Just About Cost. The Green New Deal Is Bad Environmental Policy, Too

Author's Notes:

This post may be a bit dated, but I chose it for its completeness regarding the content that, while dated, still adheres to this book's intent to point out deficiencies in the NGD proposals, whether in 2019 or 2024. I compliment the author.

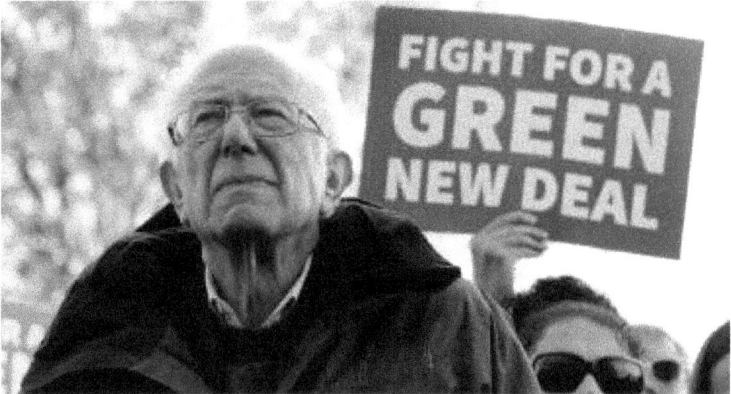

Sen. Bernie Sanders (I-VT) attends a news conference to introduce legislation to transform public housing as part of the Green New Deal outside the U.S. Capitol on November 14, 2019. Chip Somodevilla/Staff/Getty Images

Key Takeaways

- Researchers estimate it would take more than 5 trillion dollars to switch from coal, nuclear, and natural gas to 100% renewables.

- The Green New Deal would massively expand the size and scope of the federal government's control over activities, best left to the private sector.

- The reality is that environmental policies aren't good for the environment if they're so bad for people.

We're not hearing much about the 'Green New Deal' these days, but it's still a priority for some candidates, as anyone who's attended a recent Bernie Sanders rally can attest.

Criticism of the GND tends to center on cost, and rightly so. It would be extremely expensive. Researchers estimate it would take more than 5 trillion dollars to switch from coal, nuclear, and natural gas to 100% renewables, which would prove ineffective in comparison with fossil fuels.

But even if you set economic concerns aside, an ironic fact remains: In the United States and worldwide, the central -planning policies at the heart of the GND have a horrible record for the environment.

Governments in countries such as Venezuela and China (or in the past like the Soviet Union and Cuba) either routinely mismanage and waste resources or ramp up production with little to no accountability for the environmental damage that comes with it. The absence of price signals reduces the incentive to be more efficient and do more with less.

In addition, the absence of property rights reduces the incentive to conserve and gives government-controlled industries a free pass to pollute without compensating or protecting their citizens.

The Green New Deal would massively expand the size and scope of the federal government's control over activities, best left to the private sector. It would empower the feds to change and control how people produce and consume energy, harvest crops, raise livestock, build homes, drive cars, and manufacture goods.

Secondly, the Green New Deal would result in many unintended consequences. For instance, policies that limit coal, oil, and natural gas production in the United States will not stop the global consumption of these natural resources. Production will merely shift to places where the environmental standards are not as rigorous, making the planet worse.

Moreover, it's not as if wind, solar, and battery technologies magically appear. Companies must still mine the resources, manufacture the products, and deal with the waste streams. Disposing of potentially toxic lithium-ion batteries, solar panels, or even wind turbine blades that are difficult and expensive to transport and crush at landfills presents challenges. While these are solvable problems, they're seldom discussed by GND proponents.

Massive land use changes would also be required to expand renewable power. Ben Zycher at the American Enterprise Institute estimates that the land use necessary to meet a 100% renewable target would require 115 million acres, which is 15% larger than California's land mass.

Two recent National Bureau of Economic Research papers underscore the unintended consequences of energy policy on human well-being. One found that cheaper home heating due to America's fracking revolution is averting more than 10,000 winter deaths per year. The Green New Deal would wipe all of that away and reverse course by mandating pricier energy on families.

Another paper found that the Japanese government's decision to close safely operating nuclear power plants after Fukushima increased energy prices and reduced consumption, consequently leading to increased mortalities from colder temperatures. In fact, the authors estimate that "the decision to cease nuclear production has contributed to more deaths than the accident itself."

Unintended consequences.

Another hallmark of bad environmental policy is focusing on outputs, not outcomes. According to the frequently asked questions sheet released along with the Green New Deal, it is "a massive investment in renewable energy production and would not include creating new nuclear plants."

One would think that if we only have 11 or 12 years to act on climate change, we'd want to grab the largest source of emissions-free electricity we can get. But that's not the case.

That's typical for most big-government environmental policies: they're so focused on prescriptive ways to control people's behaviors that they crowd out or ignore opportunities for innovative solutions.

The reality is that environmental policies aren't good for the environment if they're so bad for people. The costs of the GND would be devastatingly high for households. Government policies that drive up energy bills are not only very regressive, but they would also harm consumers multiple times as they pay more for food, clothes, and all of the other goods that require energy to make.

By shrinking our economy by potentially tens of trillions of dollars, the Green New Deal will cause lower levels of prosperity and fewer resources to deal with whatever environmental challenges come our way. That's a bad deal for our economy and our environment.

The Green New Deal Would Harm Americans, Not Help Them

The Green New Deal (GND), a piece of legislation proposed by Rep. Alexandria Ocasio-Cortez, gained support from a large base of the younger generation.

The Green New Deal's goal is to solve America's environmental issues.

However, as a student and a young person who will experience the negative effects of this deal should it pass, I find this proposal concerning. Its main directive is to siege control and impose progressive laws to override their desires to aid the 'climate crises.'

Many support the GND because they want to solve climate change, but multiple studies have shown it wouldn't help much. According to a study by the American Enterprise Institute, the proposal would reduce global temperatures by "0.083 to 0.173 degrees," a number "barely distinguishable from zero."

***

Author's Notes:

This article is especially important. First, it was written by students at Liberty University, our future leaders, in April 2021. Second, they refuse to accept liberal progressive theorems that the NGD is pushing versus the real issue of their stated "climate crisis." I particularly like that, from the inner sanctum of Congresswoman Cortez, her chief of staff Saikat Chakrabarti even admitted the NGD wasn't originally a climate initiative at all; it was designed to disable and change our government and economy!

I wonder what else it is. Is it time for the truth rack? I believe so. I wonder what else is under the rug.

I am getting quite annoyed at the portrayal that the NGD is accepted by the young folks when the evidence presented in this very book in several places cannot be substantiated by the NGD folks that they are there in reality. They mostly are reacting to a Unicorn dream, caring less about what is involved or the associated costs that reach over $65,000 each month for every person. They do not, as a collective group, have any desire to work, are unreliable, and have no real education on which I can base my opinion, with US test scores woefully behind many countries.

I am afraid we have a shell game situation with a big megaphone, you know, like the barkers at a carnival selling you something you might be sorry you.

We must call them out for what they are and are up to!

***

GND would also be astronomically expensive. According to a Bloomberg News study, the proposal could cost up to 93 trillion dollars for 10 years, or 65,000 dollars per family per year. That's more than three times the national debt. Between paying for my tuition and student loans, I can't afford to take on more financial burden, and neither can most other college students.

In addition, the Heritage Foundation reports that the GND would increase the average household's electricity cost by about 12-14%. An economic recession or not is an additional hardship that struggling Americans cannot afford.

Another concern is that GND's initial goal was not to solve environmental issues but to restructure the economy.

I repeat, this is critically important. Rep. Ocasio-Cortez's former chief of staff, Saikat Chakrabarti, even said, **"The interesting thing about the Green New Deal is it wasn't originally a climate thing at all,"** and **"…we think of it as a how-do-you-change-the-entire-economy thing,"** according to the Washington Post.

This legislation seems to be a proposal for economic change camouflaged as a piece of environmental legislation.

Environmental policies implemented by the government from the top down, such as the GND, can lead to more pollution, which contradicts the solution this policy aims to provide. As reported by the journal Global Environmental Change, Russia (formerly the

USSR when these policies were implemented) already implemented top-down policies like the GND. However, its air quality was 1.5 times dirtier than the USA's per unit of GDP in the 1980s.

The Green New Deal aims to restructure the economy under the guise of environmental solutions.

Lastly, GND would largely expand the use of wind and solar energy, but to make this work on a national, industrial scale, it would mean clearing hundreds of thousands of square miles of forest and habitat to make way for those facilities, according to the Committee for Constructive Tomorrow. To promote conservation, habitats for wildlife would be destroyed while making room for these facilities, such as solar panel farms.

The Green New Deal is a harmful environmental policy that aims to implement faulty solutions to the environmental problem that would harm the American people more than it would help.

As a college student and lover of the environment, I strongly oppose the Green New Deal.

Comments By Readers:

- Martha Bullinger: Good points and professionally written. Good for Julia!
- D Heath: Exceptionally good article and the facts are genuinely concerning; attorneys that the facts that the items wrapped up in the GND were originally posed as how to change the economy, not global warming.

Green New Deal or No Deal

Is the Green New Deal justifiably bold or unnecessarily expansive? Do its economic prescriptions go too far or not far enough?

Dear Joe,

I want to start this debate by recognizing what we agree upon: climate change is real, humans are its leading cause, and it threatens the habitability of our planet. I, for one, am not interested in wasting time debating any of these facts. It's time to figure out what to do regarding climate change.

We know the impacts of climate change: dangerous storms, increased droughts, freshwater loss, and flooding of coastal cities, for instance. Let's focus on how we'll prevent these threats from fruition by discussing the most expansive and hopeful plan in history to fight climate change: the Green New Deal.

In its most recent form, the Green New Deal is a resolution introduced by Rep. Alexandria Ocasio-Cortez (D-NY) and Sen. Ed Markey (D-MA) that offers a set of policy directions to respond to the climate crisis and its intersection with other issues, like the economy.

The Green New Deal is a *resolution* designed to allow members of Congress to state their intent. No one designed this resolution as a policy. If passed, it wouldn't become law. It is not binding. It is a powerful step toward addressing the massive, complicated, and convoluted climate change crisis.

I recommend reading the resolution—or even skimming—if you haven't. It's surprisingly short. Go ahead; I'll leave it right here. The *Vox* explainer article on the Green New Deal is also good but a longer read. It divides the resolution into three categories that build on each other, so I'll tackle them one by one.

1. Decarbonization of the Economy

The Green New Deal (GND) calls for removing and replacing carbon-based fuels with non-carbon-based fuels. This is important because the first step in stemming the damage of climate change is slowing and eventually halting the spewing of greenhouse gases into the atmosphere. Without such action, everything else is negligible.

The Green New Deal proposes many ways to decarbonize our infrastructure, including retrofitting, reducing (not eliminating) air travel, increasing food efficiency, and transitioning to clean energy. This is radical stuff, I know.

Critics say that decarbonization is impossible. But why must our economy run on carbon? The Industrial Revolution ran on coal, oil, gas, and charcoal because those were the energy sources available at the time. Now, we know how to capture electricity from the sun and convert it into heat. We know how to capture ocean energy and convert it into light.

Carbon's status as our current energy system does not indicate that carbon-based fuels are the basis of a modern energy system. Our carbon-based energy system is causing planetary destruction that will eventually threaten the economy's function. So, maintaining a carbon-based economy will also destroy our economy (but more on this later).

Decarbonizing our economy is a noble goal, but not sufficient. That's why the Green New Deal doesn't stop at decarbonization.

2. Federal Jobs Guarantee and Public Investment

The second category of GND resolution is a federal jobs guarantee and public investment. Yes, the boogeyman of the Right: government spending. In the past few decades, some people have attempted to hijack this debate by demonizing government spending - but we need the government to spend money, especially to confront crises. Recall that President Roosevelt's New Deal and massive wartime spending during World War II got the United States out of the Great Depression. Governments need to spend money to mollify or reduce the severity of risks to the public.

But here's the big question you 've been waiting for: "How will we pay for it?"

For one, government spending stimulates the economy, often creating enough growth to result in sufficient tax revenue for the spending to pay for itself. If one wishes to offset these costs, we could increase taxes on the wealthy - recall that after WWII, we taxed the rich at over 80%, and our economy did quite well - or cut billions of dollars in unnecessary military spending.

Now that we have our cash, I'll give you two hypothetical situations.

Number one: Climate change grows out of control, coastal cities flood and force people to evacuate, the cost of natural disasters skyrockets, increased heat and pollution along with disappearing freshwater stifle agriculture, healthcare costs resultantly increase, and the economy tailspins into chaos, leading to a Great Depression-like crisis that requires massive government spending to overcome. Sounds farfetched? One of the biggest factors contributing to the sustained severity of the Great Depression was the Dust Bowl, among the nation's greatest environmental disasters. Environmental disasters ruin the economy - and climate change is one hell of an environmental disaster.

Here's the second scenario: We enact bold policies such as a Green New Deal, mobilizing millions of Americans into professions that reverse the crisis, shaping a new economy that operates without pollution or fossil fuels, strengthening our ecosystems and thus leading to better regulation against natural disasters and increased benefits from fresh water and clean air, allowing the population and economy to grow. Government spending may be high, but it averts many looming disasters and improves our current system.

Which option would you choose?

A possible retort to my scenarios may ask, "Why only two opposite choices? Can't the solution land somewhere in the middle?" Yes, it can. It depends on how much destruction we want to avoid and how we choose to do so. But we need to act, and that's what the Green New Deal calls for. It's a statement of intention to mobilize our government, citizenry, and resources to avert climate change.

So, how do we ensure that happens?

3.   Just (Fair) Economic Transition

Let's return briefly to the federal jobs guarantee. While I can already hear Fox and Friends creating far-fetched nonsense about this, it's not a radical idea: everyone who wants to work will have a job. Isn't that a basic assumption of capitalism? If you want to work, you can find a job. Top-down communist authoritarianism mandates a job for everyone, whether people like it or not. A job guarantee gives everyone the option - it seems almost... meritocratic.

And that's how the third part of the Green New Deal develops: a just transition. The Green New Deal includes some of the most robust worker protections proposed in the past few decades. Our economy needs a massive shift in a new direction to avoid cataclysmic planetary catastrophe - but we must ensure we don't leave people behind. No one would close coal factories and say, "Good luck out there!" Instead, this plan calls for retraining and reorganizing.

Things will get worse before they get better. So, to curb human suffering, the Green New Deal proposes a series of goals related to health care and living wages. It proposes directing resources to communities affected most by environmental devastation. Critics call it a laundry list of progressive ideas, but that's where they miss the point: Climate change might affect every part of our lives. Fighting climate change requires protecting aspects of our society that don't immediately seem impacted.

I'll address the last big argument we've heard: "It's not technologically feasible." I would note that the resolution often caveats proposals by specifying "as much as technologically feasible." However, to the broader point, a recent New Yorker article points out that the GND might only be off by five years or so, with most of its proposals possibly by 2035 rather than 2030. Five years? Not bad.

Researchers will have different estimates of what is technologically feasible, but that is the point. We have many policy points to argue, ideas to suggest and rework, and hours of physical labor. We have so much to learn and so much to teach.

That's the best part of the Green New Deal. It calls for a time of collaboration, innovation, knowledge sharing, and capacity building. It calls us to get out into our communities and interact with people we haven't yet interacted with, to be critical, and to idea. It's broad, sweeping, and waiting for us to fill in the details.

It will be so much fun.

Sincerely,

Ethan

***

Ethan, Thank you for your opener. Although I disagree with you on this particular debate, I very much respect your passion for the environment and climate change.

I pointed out recently that we should not be afraid to identify the points of departure in our political debates. As such, I must point out that apart from our agreement that climate change is happening and that humans are the principal cause, I have some objections to your claims that climate change "threatens the habitability of our planet." As you mentioned, this is not meant to be a first-order debate on climate change, but I think the differences in our projections may impact our prescriptions.

In general, I do not believe there is a specific date or temperature rise that will be "catastrophic," nor am I confident in our ability to predict that date or temperature, given the issue's complexity (in the non-linear sense). Additionally, many such predictions assume the continuation or exacerbation of current trends, whereas reversing such trends (through carbon capture, for example) is quite possible. Regardless, I agree that the threat of climate change and the risk of inaction is unacceptably high. So, the question has to be, what do we do about it?

In your opener, you discuss your support for the Green New Deal (GND). I will respond to your arguments momentarily, but I first wanted to explain my skepticism of any government-oriented climate action such as GND. Any government-oriented climate proposal must compete with (i.e., offset) increases in emissions over the next few decades from population growth and economic development. Most plans, including the GND, do not pass this test.

The world population is predicted to increase by nearly 2 billion (to a total of approximately 10 billion) by 2050. That is a 25% increase in the world population in just 30 years. These 2 billion people will drastically increase demand, and thus emissions, in agriculture, manufacturing, and transportation. In addition to increases in the number of people, emissions per person will increase due to international economic development. As people in the third world develop economically and start to live healthier, safer, more comfortable, and longer lives - goals that are desirable but necessary and humane - they will produce more emissions per person, as developed countries tend to do.

In summary, emissions will increase across the 10 billion humans on planet Earth. How can any policy proposal account for this? The Green New Deal proposes net zero emissions by 2050. However, most of its policies can only impact the United States, which, although a disproportionate producer, only accounts for 15% of world emissions.

Okay. Now, on to the Green New Deal itself. Greg Ip highlights my major objection to the resolution (not legislation, as you aptly point out). Essentially, the Green New Deal is two deals wrapped into one. It aims to combat global warming and also create millions of well-paid jobs for targeted groups. In Ip's assessment, with which I agree, it is "more likely to set back than advance" the climate cause because of this conflation. I would like to highlight some sections to illustrate this point. GND proposes:

- "Repairing historic oppression of Indigenous peoples, communities of color and migrant communities."

- "Guaranteeing a job with a family-sustaining wage, adequate family and medical leave, paid vacations, and retirement security."
- "Strengthening and protecting the right of all workers to organize, unionize, and collectively bargain."
- "Providing all people of the United States with (i) high-quality health care; (ii) affordable, safe, and adequate housing."

We can debate whether or not these are worthwhile policies (and I suspect we may disagree), but I hope that I can convince you that, for the sake of fighting climate change, these demands must be dropped from the Green New Deal. These radical economic proposals are not even accepted in the mainstream of the Democratic Party, let alone among Independents and Republicans. If there is to be any substantial action on climate change, we cannot conflate it with the wish list of the far Left. You claim that "fighting climate change requires protecting aspects of our society that don't seem impacted by climate change." Is this true? If you could pass policies that would reduce emissions without the progressive economic proposals listed above, wouldn't you? Since you view this issue as threatening the planet's habitability, I hope your answer is yes.

For me, the question of how we pay for it remains. I find your economic analysis to be quite cavalier. Raising taxes on the wealthy and reducing defense spending are not decisions without consequences (such as capital flight, economic stagnation, and potential reductions in national security). Additionally, we are talking about a hefty price tag here. Ignoring the social programs in the GND (i.e., Jobs and Medicare for All, which some estimates project to cost more than 50+ trillion dollars), the proposals related to climate in the deal would cost tens of trillions of dollars, according to most estimates. The Federal Government spent 4 trillion dollars in 2018 (with Medicare, Social Security, debt payments, and discretionary spending accounting for approximately one quarter each), so there is no chump change.

What is my proposal, then, you might be wondering? Though I am often a governmental cynic, I am a techno-optimist. I believe in the power of technology to build a better world, and I recognize the significant role that government plays in the preliminary stages of technological development. (You disparage the amount of money that we spend on the military, but did you know that the Department of Defense is responsible for the invention of the Internet, GPS, voice recognition, virtual reality, and many other breakthrough technologies, several of which are being used in the fight against climate change?) The United States of America has unprecedented technological research, development, and commercialization prowess. We need to harness this power toward the challenges that climate change presents, along with market-driven incentives and policies, to build cost-effective technologies that are economically competitive while reducing and removing emissions. Don't just take my word for it. Listen to Bill Gates' thoughts on the topic.

To begin to solve the climate change problem, as with all major problems facing our country, we cannot steamroll the opposite party and force our legislation down their throats. The GND would require Democrats - and progressive Democrats at that, given the four "No" votes in the Senate by Democrats on the GND - to control the White House, the House of Representatives, and sixty seats in the Senate. This seems quite unlikely; as you mentioned, we don't have time to waste. We need to come together and identify bold, bipartisan steps to take. We need to pass legislation that can withstand administrations. I am afraid that the Green New Deal fails this test. In my view, we need to go back to the drawing board.

Best,

Joe

***

Hire Hollywood to collaborate with the creators of AVATAR, the fictional movie about an avatar who controls Air, Water, Fire, and Earth. Just think about how much money we could save.

— a suggestion for Washington by Ethan Freedman

What Is the Green New Deal?

Or should I ask *what the new, new Green Deal is?*

They seem to push out a lot of them, don't they? Lots of Washington fluff, assigning billions and trillions of expenses to address too many areas (like they'll ever get them off the ground or accomplished)! Just had to chime in to let you know I'm still here. More good stuff to come —stay with us.

The term "Green New Deal" has been used to describe policies that address climate change policies and investment in renewable energy. The United Nations (UN) announced a Global Green New Deal in 2008. Former President Obama added one to his platform when he ran for election in 2008, and Green Party candidates, such as Jill Stein and Howie Hawkins, did the same.

A Call to End Fossil Fuels and Build Green Jobs

By **Deborah** D'Souza2/12 2024Reviewed by: **Cierra Murry**

Fact checked by: **Kirsten Rohrs Schmitt**

Thank you Deborah, Cierra and Kirsten for your contribution.

I am totally confused over using a bill or one of several resolutions that can be by one or both houses of our government. The funds delivered in these NGD resolutions are not minor sums of our tax money, and I am concerned about proper monitoring these funds.

Let me give you an example of one: we got shafted for something the government does not want to clean up, and the government assumes no liability for a bankrupt company; the company is Solyndra. This other company, which just declared insolvency and bankruptcy, might cost us trillions to clean it up, and it is also concerning because it was funded under the NGD monies and produced solar panels widely used by homeowners whose warranties are no longer good. You will find more details on this later on. What you will see ahead is indeed scary. This raises the question: How many other questionable NGD situations exist?

I live in Tennessee, a state that has had its very own Trail of Tears for a disastrous Native march to a place where they would be "out of the way" and be taken care of. Right! This is obviously socialist driven. Yes, it does look great with lots and lots of promises but lots of obvious potential pitfalls that will cost us dearly; it sounds familiar.

Have you ever felt overwhelmed by the massive cash outflow on NGD proposals, bills, and literature? While certain things are proposed, they often seem like "pie in the sky." Obvious jobs never materialize, making us stuck with bad information about climate data—something we have no control over but need immediate funding for while the list of challenges continues to grow.

I have amassed many informational sources related to you, and what I have already posted is a prelude to the actual effects of the NGD you absolutely need to know. I thank those who have already contributed to their articles.

In 2019, a Green New Deal was a congressional resolution introduced by Representative Alexandria Ocasio-Cortez of New York and Senator Edward J. Markey of Massachusetts.

Key Takeaways

- The term "Green New Deal" has been used to describe policies that aim to make systemic change.
- The term was first used by Pulitzer Prize-winner Thomas Friedman in January 2007 and made popular by the proposal of Rep. Alexandria Ocasio-Cortez and Sen. Ed Markey in Congress in 2019.
- The deal emphasizes environmental and social justice while calling for new job creation.

History of the Green New Deal

The term "Green New Deal" was first used by Pulitzer Prize-winner Thomas Friedman in January 2007. Friedman recognized a solution to a climate that would require money and effort and upset the fossil fuel industry. Transitioning away from fossil fuels, he argued in a New York Times column, would require the government to raise prices, introduce higher energy standards, and undertake a massive industrial project to scale up green technology.

"The right rallying call is for a 'Green New Deal,'" he wrote, referencing former President Franklin D. Roosevelt's domestic programs to rescue the country from the Great Depression. "If you have put a windmill in your yard or solar panels on your roof, bless your heart. But we will only green the world when we change the nature of the electricity grid—moving it away from dirty coal or oil to clean coal and renewables.

Proposals 2019-2024

In 2019, Representative Alexandria Ocasio-Cortez and Sen. Ed Markey introduced a 14-page, nonbinding resolution in Congress calling for the federal government to create a Green New Deal. The resolution had over 100 co-sponsors in Congress and attracted several Democratic presidential candidates. According to the proposal, the U.S. and its historical contribution to greenhouse gas emissions were disproportionate.

On March 26, 2019, lawmakers in the Senate voted 57-0 against advancing the resolution, with 43 out of 47 Democrats voting "present" to avoid a formal position. The Green New Deal proposed tackling the climate crisis with a 10-year mobilization and addressing an issue that 60% of Americans said already affected their local community. The deal promised to tackle economic inequality by creating high-quality union jobs and called for protecting workers' rights, community ownership, universal healthcare, and a job guarantee.

In April 2023, Senator Markey and Congresswoman Ocasio-Cortez reintroduced the Green New Deal Resolution, but the resolution has not yet passed as of 2024. However, since the Green New Deal was first introduced, supporters saw some federal investment to fight climate change included in the 2022 Inflation Reduction Act. I hope the readers keep track of the proposed adjunct bills to keep the NGD funds flowing. The Act thus far has cost us dearly, and jobs are victims already.

In 2024, five years after the introduction of the Green New Deal, polling from Data for Progress shows that 49% of voters view climate change more seriously than in the past.

Goals of the Green New Deal

The plan's main goal is to bring U.S. greenhouse gas emissions down to net zero and meet 100% of power demand in the country through clean, renewable, and zero-emission energy sources by 2030. The proposal calls for job creation, along with access to nature, clean air and water, healthy food, a sustainable environment, and community resiliency by:

- Providing investments and leveraging funding to help communities affected by climate change
- Repairing and upgrading existing infrastructure to withstand extreme weather and ensuring all bills related to infrastructure in Congress address climate change.
- Investing in renewable power sources
- Investing in manufacturing and industry to spur growth in the use of clean energy
- Building or upgrading to energy-efficient, distributed, and smart power grids that provide affordable electricity.
- Upgrading all existing buildings and building new ones so that they achieve maximum energy efficiency, water efficiency, safety, affordability, comfort, and durability.
- Supporting family farming, investing in sustainable farming, and building a more sustainable and equitable food system
- Investing in transportation systems, namely zero-emission vehicle infrastructure and manufacturing, public transit, and high-speed rail
- Restoring ecosystems through land preservation, afforestation, and science-based projects
- Cleaning up existing hazardous waste and abandoned sites
- Identifying unknown sources of pollution and emissions

- Collaborating with the international community on solutions and helping them achieve Green New Deals.

Author's Feedback:

I do not see how the government can objectively review and determine climate change issues for the entire USA and prioritize helping communities by obtaining, assigning, or leveraging funds for any private concern. There would be opposition based on empirical building requirements across the nation, but wait, don't these have states, and don't states have universal building codes? Yes, they do.

- Here is how the US Dept. of Transportation and states work together: The state collects taxes, has to pay the government based on data provided, and makes payments accordingly. The Government apportions monies based on data and returns funds for projects the states define on a Cherry sheet. THERE IS NO NEED TO TELL THE CONGRESS TO MENTION THE CLIMATE CHANGE.

- After all these years of pushing wind and solar, and by the way destroying wildlife and land mass us forever, only 17% of all installations since they have been promoted feed energy grids. I can see innovative water reclamation and hydro energy wherever practical, but do not destroy my area!

- Yes, you do. I am now working with a three-county agent in TN, and this is already being done. They are funded in different discussions. No bucks are needed here.

- If I were the government and specifically the NGD folks, this is a hot potato with the Solyndra fiasco and now a possible Trillion-dollar bankruptcy just four years ago the NGD funded. This affects millions of people. It seems whatever they touch turns to, well, you know.

- Do not squander millions of acres on unreliable (read fair weather solutions) that are ineffective for stable grid use. How many years now? And in 2005, the yields were just 5%, and now, 20 years later, just 17%. You must look at reliable generators to feed a growing population's needs.

- Now, I particularly dislike this one. All states have building code enforcement agencies. In fact, I am currently working with an agency that is holding feet to the fire for an issue I am having with a state-licensed building corporation. Their code is solid, and if your changes fit their perceived needs, what you want might work. Again, no money is needed from NGD magic bucks.

- I do not see an issue with this, and the message could be PSAs on public television if The Pillow guy will get off all stations and let someone else say something. Say, don't you have an in on PBS?

- Yes, Transportation: the US ranks 10th worldwide for rail access and use.

- Let the states take care of this issue. We do not need government assistance. It is said, "Everything they touch turns to -well, you know.

- It's a government job; say what they didn't and won't clean up Solyndra. Who can we trust?

- The government clean-up of hazardous sites; this is the EPA's job, I thought, and having been involved with the country's 3rd worst Superfund site, it did an excellent job, including making it a brownfields site for my town. Let them do their job. Hey, BTW, what happened to the Transportation Secretary and the poor city in Pennsylvania? Did he ever get there? Just asking for a friend.

This one I have very much dislike for. Overreach into other countries by the USA in this so-called climate issue is concerned! Take care of us first and last. Stop spending like a bar hobo!

Rank	Country/Region	Passenger-kilometres (billions)	Data year
1	China	1,550[2]	2019
2	India	1,157[3]	2019
3	Japan	446.7[4]	2019
4	Russia	133.6[4]	2019
5	France	112.6[4]	2019
6	Germany	100[4]	2019
7	South Korea	93.9[4]	2019
9	Italy	56.6[4]	2019
10	United States	32.5[4]	2019
11	Ukraine	28.4	2019
12	Spain	27.3[4]	2019
14	Poland	22.1[4]	2019

There's not much I know about the Billion-dollar gift that Pelosi and Newsom got. I am sure most of my readers are aware of it. Did we get our money back? I like the idea of electric buses, but I am not a fan of electric cars and buses due to their battery issues and the care they require. When it comes to trucks, it's a losing proposition, but if you can't ship by rail, they are not very capable, it seems. There will be a massive loss of trucking jobs because of that if they insist on rail.

Has anyone figured out how we are going to cycle the unrepairable (not burnt) car batteries, which are said to be only good for a brief time? Are we going to have another EKK? What about the trash problem for the groundwater as we have now with solar panels and old windmill blades? Oh, wait, we send the bad ones to third-world countries to use. I wonder what they are doing with them. I am in favor of the new hydrogen motors. Worked on solving the problem in the late '70s, we solved the venturi problem., sold it to Ford, and "into the darkness."

Climate Change

A common rebuttal from supporters of the New Green Deal to its opponents is that, although it will be expensive to implement, not doing so will be more costly eventually. To stop temperatures from rising beyond 1.5 degrees Celsius—the target aimed for in the 2015 Paris Agreement—global emissions need to reach zero by 2050.

The federal government spent $450 billion due to extreme weather and fire events between 2005 and 2018, according to a 2018 report by the U.S. Government Accounting Office. In 2022, the U.S. experienced 18 weather and climate disasters, each costing at least $1 billion.

According to a 2019 study, climate change is projected to cause more than $500 billion in economic loss in the United States alone each year by 2090.14 Independent research shows that about 10% of the global economy's value could be wiped out by 2050 if temperatures continue to rise by 3.2 degrees Celsius and the world fails to meet the net-zero targets outlined by the Paris Agreement.

Supporters

Advocates of the Green New Deal who promote a heterodox macroeconomic framework called Modern Monetary Theory (MMT) believe the government should not be overly concerned about the cost. "The federal government can spend money on public priorities without raising revenue, and it won't wreck the nation's economy to do so," a group of prominent MMT supporters wrote in an op-ed for The Huffington Post.

Author note: This author believes in conservative spending and once you approach the level of over spending, meaning calculated income, is socialist and leads to spending "funny money" you stop spending. This MMT theory pushed us over the line with RESOLUTION draws of things that were spent offshore, where things industry could do, and did, failed solar, wind power and spending reward money to people who would bend to their exorbitant projects that failed or were marginal. The damage they caused was incalculable and its impact will be with us for decades.

The Green Party, whose plan also calls for America to move to 100% clean energy by 2030 and a job guarantee, says it will result in healthcare savings and military savings. The party's deal also advocates for a robust carbon fee program. Healthcare and other savings were also touted in a 2015 study by a group of scientists from Stanford University and the University of California, Berkeley, which said the U.S. can replace 80% to 85% of the existing energy systems with ones powered entirely by wind, water, and sunlight by 2030 and 100% by 2050.

However, as a critic of this position, I find these claims unrealistic. It is not feasible to construct the necessary infrastructure to replace 83% of fossil fuel-based power sources within the suggested timeframe. To ensure their viability, reputable analysts should reality-check these projections.

Critics

Critics call the Green New Deal socialist, extreme, or impractical. Some have warned that environmentalists "want to take away your hamburgers. The kind of sweeping overhaul the deal proposes would be expensive and require significant government intervention. The center-right American Action Forum pegs the maximum cost at 93 trillion. At the same time, Tax Policy Center senior fellow Howard Gleckman said the plan could slow the economy by increasing national debt and even driving jobs overseas.

"Instead of the Green New Deal, the federal government could adopt a revenue-neutral carbon tax to decrease emissions without exacerbating the fiscal imbalance," said Jeffrey Miron, director of economic studies at the right-wing Cato Institute. Edward B. Barbier, the American economics professor who authored the report that formed the basis of the UN's Green New Deal, said that, instead of deficit funding, the government should use revenues from dismantled subsidies and environmental taxes.

How Will the Green New Deal Affect Investors?

Investment opportunities may arise with the passage of a Green New Deal. According to UBS, there is a long-term trend towards more sustainable and green ways of producing and consuming, with investments in environmentally oriented sustainable industries.23. While a Green New Deal doesn't explicitly call for eliminating fossil fuel usage, it may hit the industry hard, including nuclear energy stocks.

Who Wrote the Green New Deal?

Thomas Friedman first introduced the idea of a Green New Deal, which recognized that there was no easy solution to climate change. In 2019, Rep. Alexandria Ocasio-Cortez (D-NY) and Sen. Ed Markey (D-Mass) introduced a proposal for a Green New Deal in Congress, though it failed to pass in the Senate. The proposal was reintroduced in 2023.

Mr. Friedman obviously did not follow his own advice "no easy solution" but passed it onto people who politically embellished and drove a possible good thing into the ground to really make it impossible to be successful and a huge "money sink" that funded projects that failed, failed to live up to the hype and a huge climate and jobs killer. Can't forget the green projects that are all underperforming, severely affecting the environment and the asinine assumptions that all electric be "the end all be all" solution. I could go on, but the failure list would add to many pages to this book.

How Would the Green New Deal Create Jobs?

The Green New Deal promises to create millions of jobs by shifting money from the fossil fuel industry to green technology.

Is the Green New Deal socially and economically viable?

Adam Roach

As written, No.

This is not a Bill that was written to pass; it was written to make noise and fire up kids for politics. In that arena, it looks like success.

The first challenge is the timeline. It would take 12 years to alter the US transportation and energy infrastructures completely. That's a tall order, but it's not insane since both infrastructures need serious updates. Our Energy Grid is an antique and apparently vulnerable to physical terrorism and cyber-attacks from well-equipped hackers. Meanwhile, our transportation infrastructure seems ten years behind wherever it should be.

I'm going to consider socially viable as to whether the Green New Deal would lead to large-scale social unrest, at least if implemented as suggested.

So, the answer to both is neither socially viable nor economically viable.

This is not due to what it wants to accomplish (0 net emissions and a transition off of fossil fuels) but how it proposes to complete that, particularly given the ambitious timeline and the complexity of overhauling entire sectors of the economy like energy and transportation.

Is the Green New Deal" realistic?

Douglas Eckberg

No.

The name of the GND sounds as though it is focused on climate change, which is to say, energy production and use. It would be hard to oppose if — like me — one is seriously concerned about global warming and is taken with the urgency of the most recent IPCC report (Summary for Policymakers of IPCC Special Report on Global Warming of 1.5°C approved by governments). But while that is part of it, it is a grab-bag list of things a lot of Democrats would like to have the government do. That doesn't make it bad, though one person's list will be different.

Do you think the Green New Deal is realistic?

Susannah Moyer

Not only is it unrealistic, but it's also stupid.

The easy part is to ask what part of Medicare for All and Free College is green. What part of paying people more or paying people not to work is green?

This proposal sounds more like an aggressive socialist proposal.

Getting rid of air travel

How will the US be even more competitive by eliminating air travel? What would Al Gore do without his private jet, the Clintons, Bezos, or Gates?

California couldn't even build a high-speed rail from Los Angeles to San Francisco. At the start, the project was budgeted to 4 billion. It has since faced numerous delays, cost overruns, and logistical challenges, raising doubts about the feasibility of such large-scale infrastructure projects under similar timelines.

Is the Green New Deal socially and economically viable?

Let's have a look!

This is all about CO2 - plants naturally convert CO2 into growth and oxygen.

The Green New Deal prohibits diesel or gas motors, so **diesel or** gas "charging/generator" motors are out.

So, you know, that also kills most of the locomotive power for rail traffic since they burn lots of diesel-generating power for those electric motor systems.

Currently, the infrastructure isn't in place, the charging ability is not there, the technicians are not there for the motors, the power grid is not there, and there is no network of charging system.

Do you think the Green New Deal is realistic?

Cary Aguillard

Most of Trump's tweets are not worth reading, but his sarcastic tweet in response to the New Green Deal sums it up well.

"I think it is very important for the Democrats to press forward with their Green New Deal. It would be great for the so-called "Carbon Footprint" to permanently eliminate all Planes, Cars, Cows, Oil, gas, and the Military - even if no other country would do the same. Brilliant!"

The proposed program would radically limit and revise each item he listed. Picture America with a reduction or limitation of auto manufacturing, airline manufacturing, and fossil fuel production, all of which would undergo drastic cuts or shifts to align with the Green New Deal's goals of net-zero emissions and a transition to renewable energy sources.

Do you agree that the new Green Deal is good or bad for America? Why?

Howard Robinson

The New Green Deal is based on a fallacy that Apocalyptic anthropogenic global warming is a reality, which, of course, it isn't. The main purposes are to usurp freedom from individuals, give it to the government, and destroy the middle class by increasing the cost of energy and using the increased expenses to subsidize or provide it for free to the lower class. Only the elite, including the wealthy and politicians, will be exempt from its rules. It will weaken our country and destroy the economy to destroy America. It is evident to me that anyone who actually believes this nonsense hates America.

Why is the Green New Deal hard to ratify?

<u>Dirk Sayers</u>

I think the short answer to your question is that the Green New Deal is an ambitious, aspirational document suffering from (at least) three deficiencies:

1. It's *really* ambitious (read that probably overly optimistic)
2. It lacks *actional specificity*.
3. It threatens to break *a lot* of rice bowls, beginning most notably with the carbon-based energy interests.

If you haven't read the Green New Deal resolutions yet, I commend it to everyone. What's important to recognize is that it's a resolution rather than a bill in the strictest sense of the word. There's a lot of entrenched resistance to it, some of it justified, a lot of it rooted in misunderstandings or ideological opposition to the board changes it advocates for. While the Green New Deal outlines ambitious goals, its real purpose may be to spark discussion and push for action on climate change rather than being a direct, executable plan.

Why should I support the Green New Deal?

<u>Paul Masters</u>

I would avoid it. I will be charitable with it but suffice it to say it seeks to employ highly questionable economic strategies to achieve an unrealistic goal.

The Green New Deal is composed of two separate things. First is what it wants to do, which is transition to a 100% renewable energy grid in 10 years. The second is how it wants to do it, which is the nationalization of much of the economy for a government-led infrastructure project.

The goal is impossible with the strategies proscribed, and the how is asking for disaster, exacerbated by the goal it is impossible.

<u>Mike Jones</u> ·

Is Alexandra Ocasio-Cortez's "Green New Deal" viable?

This is not, at this point, a question that has a rational answer.

At this point, what has been put forward is a statement of direction, a set of goals turned into a "sense of Congress" resolution, not a set of specific proposals ready to become law.

It's a starting point for discussion, meant to stake out a "left" position and start a discussion. Conservatives, for the most part, seem to be instead reacting by attacking both the proposal and AOC as everything from "unrealistic" to "insane", which I strongly suspect is because they know that many of the ideas in the initial statement are broadly popular.

What do you think of the Green New Deal?

<u>Jon Lazal</u>

<u>The Green New Deal:</u>

- Vague and poorly thought out.
- Literally has a clause for socialism ("public ownership")
- Unrealistic goals
- Unrealistic promises
- Costs trillions of dollars
- Isn't economically feasible?
- Used to have a line about providing economic security for those UNWILLING to work in the earlier drafts.
- Arrogant in its nature
- Ideologically driven.
- Tries to solve every other controversial social issue while being masked under climate change.

- The entire bill is less than 2000 words.
- Not even hard democrats are willing to support it.
- The person behind the bill is economically and politically illiterate.

What's controversial about the Green New Deal?

Bill Cravens

Where shall we begin?

Those promoting it put an insane level of confidence in "renewable, non-polluting energy sources." Wind and solar power are fine, and they can "augment" our other sources of energy, thus taking a little bit of the burden off of the environment. But there simply is not enough available, nor are they dependable enough, to even begin meeting the needs of 21st-century technical civilization—not by a long shot.

2) Conversion to electricity does convey some benefits, in that it generates pollution at power plants, rather than through ten million engines. But it still has issues.

Is "clean energy" economically viable?

Frank Zucco

Where have you been?

Not only is clean energy economical, but it is also the ONLY energy that is economical today. The change happened in most places in 2017—almost six years ago.

Oil and gas, nuclear, and coal are getting rapidly too expensive to continue using as we do.

Diesel fuel, Aviation Kerosene, and Fuel Oil—the mid-fraction distillates—are in shorter supply each year. Tar sands and fracking cannot replace them. These are critical fuels for agriculture, waterborne shipping, trucking, rail transport, and aviation, and they are needed just to get the oil out of the ground, refine it, and deliver it to market.

Can the Green New Deal deliver on its promises?

Timothy X Rowe

The Urine thing is yet another conservative LIE!

It's really not meant to deliver on anything. I've seen this kind of thing before, and it has its roots in extreme advocacy and how to implement change. This is not a criticism or negative statement against environmentalists. Still, sometimes, "true believers" and folks who get seriously worked up about the potential risks of climate change believe that drastic action is required.

For your information, climate change refers to global warming until the warming stops and the Earth starts cooling.

Is public opinion divided on the Green New Deal?

Bob Watts

Sure, the public is divided on Alexandria Occasionally-Coherent constant ranting and raving, speaking incessant psychobabble that only an uneducated fool would consider newsworthy.

I believe that the majority of Americans want to invest in renewable energy. It looks good on paper, maybe the answer, and deserves honest consideration.

HONEST CONSIDERATION is NOT what Old Demented Joe, Kacklin' Kalamity, and the rest of the so-called 'progressives' in the Democrat party are doing.

Conservatives KNOW that we cannot just snap our fingers and say: No more fossil fuel will be produced in this country.

What are some alternatives to the Green New Deal?

Bruce Cunha

Start a "Manhattan Project" to find a non-polluting energy source. Simultaneously, significant research into better storage systems for electricity.

Remember, we have changed fuel sources at least four times in the past. Each time, the change came about when a less expensive form of energy was available, and it made money for the companies that promoted it.

All the Green New Deal ideas are not economical.

What is the economic argument behind the New Green Deal?

Charles

We are going to face catastrophic environmental and economic damage if we don't lower global warming and stop climate change from getting worse. There isn't enough money to fix these problems once they reach dangerous levels, and we are right at that point. Trillions of dollars won't fix the problems once they reach the point of no return, and our lives and our children.

Key Highlights of the Green New Deal:

Something needs to be done to help us avoid a Climate catastrophe. Air and water pollution are a threat to public health really.

The **Green New Deal**, as proposed by Ms. Cortez, is problematic for key **reasons:**

1. No car transport with gas. (or the extent to which technologically feasible). By the way, renewables only account for 17% of energy.
2. No air transport with gas. (or the extent to which it is technologically feasible)
3. Has solar or wind been deployed at scale without blackouts?
4. No beef (due to emissions)

Is Bernie Sanders's New Green Deal affordable?

Jonathan Lloyd

Absolutely, yes, it is affordable.

It had better be, or we are all screwed.

We can work to fix global warming, or we can all twiddle our thumbs, waiting for the inevitable sea rise, famines, crop failures, mass extinctions, and mass migrations, or we can act—and spend—right now.

Spending right now will be far, far the cheaper option. So, the real question becomes:

Is waiting for the consequences of global warming affordable?

No, decidedly not.

Can you imagine how much it will cost when millions migrate away from the coasts and cities attempt to re-make (and inevitably fail to) their infrastructure?

What is the status of the Green New Deal?

Jane Smith

Pretty much dead in the water! But it won't go away for good, AOC took too far a leap, and she will be punished for it...but she did start a dialog! By setting out a ridiculously outlandish proposal she may have been indeed very strategic. If she was, well kudos to her! The context of the conversation has moved from denying global warming to just how far is too far.

Is the Green New Deal a reasonable approach?

Benjamin Johnson

No, it isn't.

The Plan for a Green New Deal (and the draft shall be developed to achieve the following goals, in each case in no longer than 10 years from the start of execution of the Plan:

Only ten years? Why?

1. 100% of national power generation from renewable sources.

Renewable sources? Does nuclear count? We could go to a 100% nuclear economy within ten years; we'd already be there if we hadn't stopped building reactors in the 70s. But since this is a Liberal, probably not unless we finally perfect fusion. Ain't gonna happen if we only have solar and wind.

Can the Green Deal actually work?

Johannes W. van der Spek

Only if you ignore science, unintended consequences, and the real world.

In a CG world, it is simple.

That is why every person confronted with reality hides behind saying it is a Resolution.

- If it is only a Resolution -
- why waste Congressional Time discussing a Resolution
- if it is non-binding - why waste Congressional time on it
- here is a novel idea - instead, Let's have Congress do something.

What is that in reality, let's take a good, hard look.

If the Green New Deal becomes law, the U S would have to go 100% green in 10 years.

Can the Green New Deal deliver on its promises?

Frank Merriam

There's an old joke, it goes: What did they use to light their homes in the USSR before candles?

The answer: Electricity

When governments try to get things done, things usually go backward.

So, the answer to whether the Green Deal can deliver is absolutely not. Not even a little bit. First off, it is a promise made in the absence

116

of technology needed to deliver. Renewable energy returns a fraction of the power per dollar spent. If it weren't for vast government incentive programs, no one would waste their time on it which is how you know it isn't likely to be profitable anytime shortly, because more people would risk if it was possible.

Marian Hubbard

Originally Answered: The Green New Deal is this socially and economically viable?

Neither.

The elephant in the discussion is that it would take a revolution to overthrow the government and bring in a socialist regime to take complete control of not just all the means of production, but every other part of American society.

And all the indications show that people in the U.S. are pretty happy with the country as it is and are optimistic about our future.

Countries just don't overthrow successful, economically strong governments.

About the Republican Policy Committee

Established in 1947, the Republican Policy Committee (RPC) is an official Senate office that serves all Republican members of the United States Senate. The Policy Committee is directed and managed by its chairman, who is a member of the Senate Republican Leadership and who is elected by his or her peers for a two-year term at the beginning of each Congress. Senator Joni Ernst of Iowa currently serves as the 17th Chairman of the Republican Policy Committee.

According to the Rules of the Senate Republican Conference:

The Policy Committee shall consider the legislative program in the Senate and the question whether any Party policy is involved, shall prepare and present recommendations for action by the Conference, and shall advise all Senators on legislative matters that they desire to present to the Policy Committee.

In the tradition of past chairmen, Chairman Ernst is committed to providing all Republican Senators a forum for substantive policy discussions and debate. This is principally conducted through the weekly Policy Lunch, which is a standing meeting of the Policy Committee that includes all Republican Senators and is typically held every Tuesday when the Senate is in session. The Policy Committee also provides Republican Senators and staff with issue research, written and in-person policy briefings, in-depth analysis of all pending legislation, and summaries of member voting records. The Policy Committee also manages Channel 5, which is available on all Senate televisions, and the Trunkline, which is the intranet for Senate Republican offices. The Policy Committee also holds weekly standing meetings of Republican legislative directors and committee staff directors to discuss the legislative agenda.

The Committee's origins date back to 1947 and its founding chairman, Senator Robert A. Taft of Ohio. The Committee was created following a bipartisan proposal by the Joint Committee on the Organization of Congress in 1946. The Committee was originally composed of nine Republican Senators. Today, official meetings of the Policy Committee include all Republicans…

December 11, 2018

Author's Note

Before posting the article below, I thought it would be prudent to provide you with some background on the RPC. Just to clarify, they haven't been idle. As early as December 2018, they were actively addressing and evaluating the NGD with concern. Their comments are included below, and any new RPC documents that address the NGD or highlight specific concerns will be posted as they become available. I realize the article is dated, but in my view, it remains highly relevant.

Green New Deal: A Crazy, Expensive Mess

Key Takeaways

- The Green New Deal proposes carbon neutrality and 100 percent renewable power within 10 years – experts say the latter goal is impossible by 2050, let alone 2029.
- All-renewable electricity would mean closing every nuclear, coal, and natural gas plant and cost $7 trillion.
- The Green New Deal's job guarantee would replace productive work with federally backed make-work jobs, distorting the labor market and hurting private businesses.

House Democrats are rallying to a proposed "Green New Deal" to make the United States carbon neutral within 10 years. While the plan is being sold as a solution to climate change and a jobs program, it would cost trillions and ignore the rest of the world's contribution to climate change.

I take great umbrage with the last two lines, especially the voiced remark that this early version is extremely revealing of the mindset concerning disregarding the rest of the world's responsibility in addressing climate change. It was, and continues to be (as of 2024), a key factor in what is now a global issue. Everyone must contribute to the effort, or, as the saying goes, "we'll all go down with the Titanic.

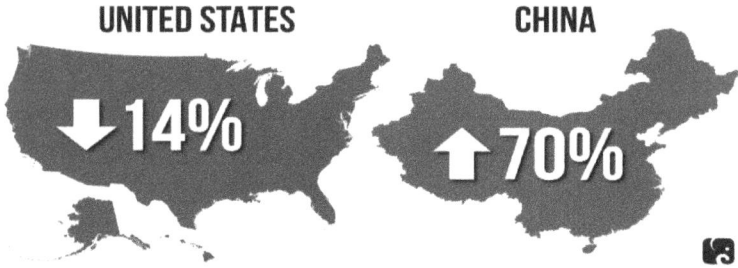

UNITED STATES ↓14%   **CHINA** ↑70%

2005-2017: U.S. Cut CO2 Emissions; China Dramatically Increased Emissions

If enacted, the Democrats' idea would require 100 percent renewable-sourced electricity and changing transportation and industry to eliminate carbon burning. While Democrats have not released a cost estimate for their proposal, independent **estimates** suggest it would cost at least $7 trillion. Even with all that spending, the international nature of climate change means that the plan could fail to affect global carbon dioxide emissions meaningfully. Twenty-two House Democrats have endorsed the Green New Deal and plan to create a select committee early next Congress to begin drafting legislation.

GND Actually Ignoring emissions trends.

Even if the Green New Deal functioned exactly as Democrats are trying to sell it, the U.S. might spend $7 trillion only to find that climate change still got worse. Global emissions could rise because the Green New Deal ignores the international climate change aspect. The United States **generates** only around 15 percent of global carbon dioxide emissions.

Of the G-20 countries, the United States has the best recent record on carbon dioxide emissions reductions. The U.S. **cut** 862 million tons from 2005 to 2017, a 14 percent decline. Over the same period, the European Union reduced its emissions by just 771 million tons, and global emissions rose 26 percent. India increased its carbon dioxide emissions by 1.3 billion tons, and China increased its emissions by 4 billion tons—a 70 percent increase.

Author's Note

This may sound petty, but it's important to recognize that this is just one story among, I am sure, millions across the USA, all following the EPA's 1978 pollution control compliance guidelines. I was a protégé of Al Marshall of Marshall's stores, and we received an order from the State that we had to install air and particulate scrubbers on our furnace, where we burned trash and boxes. The order was strong, and we could not burn until we had a scrubber installed, so all our castoff materials had to be carted to a landfill. Never thought about it until I authored this book, but I can only imagine all the pollutants from millions of stores going into the atmosphere and all the landfill gasses that landfills produce. WOW! Our government was on the job back in the late 70's.

If viewed from a global perspective, the Green New Deal is a colossal waste of money that ignores the vast majority of global and atmospheric problems and largely ignores those things it cannot control.

Renewables not yet up to the task

Producing 100 percent of electricity from renewable sources is a practical impossibility shortly. Scientists doubt it would be achievable by 2050, let alone 2029, the deadline Democrats would set. Such a massive overhaul in power generation would require the closure and replacement of about 83 percent of U.S. electricity generation, including all coal, natural gas, and nuclear plants. While nuclear energy does not release carbon dioxide, it is not renewable and would not be allowed under the Green New Deal. Today, renewable electricity — mainly wind, solar, and hydroelectric — provides only 17 percent of American electricity.

The entire grid would have to be revamped because renewable energy sources are not as stable as traditional electricity sources. Massive amounts of energy storage would be needed, requiring dramatic technological improvements. New transmission networks would be needed to bring electricity to areas less conducive to solar or wind generation. Innovative technologies would be needed to regulate electricity frequency, output fluctuations, demand response, and voltage management. Even mainstream left-leaning environmentalists have argued against a 100 percent renewable strategy because of unsolved technical problems and enormous expense relative to benefits.

Job Guarantee and A Grab-Bag of Other Left Policies

The Green New Deal also proposes a job guarantee for every American who wants one. Beyond promising a not-yet-defined living wage, the details are scarce. Other job guarantee proposals popular among Democrats generally promise federally subsidized jobs in fields such as infrastructure and public works, child care, and the environment for anyone, whether they are currently employed or unemployed.

A job guarantee is a deeply flawed policy. The only "guarantee" is a badly distorted labor market. First, it is doubtful that the government could successfully implement such a vast program. Once underway, the damage to private businesses would be devastating. The effects would fall heavily on small businesses, which could be forced to shut down if they could not match the federal wage or because workers prefer one of the "guaranteed" jobs over their current one. It could effectively nationalize a considerable portion of the economy. A job guarantee also would double down in Washington, picking economic winners and losers. Finally, the cost would be enormous. The authors of one prominent proposal peg theirs at $543 billion annually.

Democrats promise their Green New Deal will "mitigate deeply entrenched racial, regional, and gender-based inequalities in income and wealth." They add the possibility of including basic income, universal health care, and union involvement. House Democrats have yet to explain how energy policy will achieve these goals and how they relate to climate change.

Knowledge of Wharton School at the University of Pennsylvania

Whatever the merits of the "Green New Deal" that two U.S. Democrats unveiled earlier this month, it surely has raised the temperature of the debate on climate change, jobs, income inequality, health care, housing, and more. In a broad sweep at progressive initiatives, a **resolution** introduced by New York Congresswoman Alexandria Ocasio-Cortez and Massachusetts Senator Ed Markey calls most prominently for the U.S. to move off of fossil fuels and become carbon-neutral in a decade.

While supporters have hailed the move as long overdue, critics have dubbed it as sloganeering or, at best, unrealistic in its goals. Republican Senate Majority Leader Mitch McConnell has said he will bring the bill up for a vote. Still, Democrats are dismissing that move as a political ploy rather than a serious debate on the merits of the plan.

In any event, the discussion on the Green New Deal makes room for "messaging" the urgency to combat climate change and the potential for it to change electoral results, said experts at Wharton, Syracuse University, and the University of Texas at Austin. The specifics of implementing the plan are in a gray area, such as creating high-wage green sector jobs and dealing with the displacement of jobs in the fossil fuel industry. However, many elements of the plan are workable, given enough government funding and political will, the experts said.

The plan seeks to not only achieve net-zero greenhouse gas emissions within 10 years; it also wants to create "millions of good, high-wage jobs," boost infrastructure investments, provide clean air and water, healthy food, and access to nature, and prevent the oppression of marginalized communities, among other goals.

The Green New Deal is "not a detailed plan that gives some specific outcomes, [but] a framework proposal ... and is a good idea," according to Eric Orts, Wharton professor of legal studies and business ethics. Orts is also director of Wharton's Initiative for Global Environmental Leadership. He noted that it is a "non-binding resolution," which means it doesn't intend to secure votes in Congress on its proposals but aims to air opinions and create public and political discourse.

David Spence, professor of business, government, and society at the University of Texas at Austin's McCombs School of Business and a chair in law at the university's law school, termed the Green New Deal as not "a plan [but] a statement of goals." He said that even as it "aims at a rapid transition to a cleaner energy mix," it is "big and broad" and "very ambitious."

Spence said there would be obstacles for the Green New Deal in both securing political support and implementing specific proposals, such as seamlessly replacing fossil-fuel jobs with high-wage "green" jobs in construction and renewable energy. "There are a lot of questions about how you get from A to Z in each of these questions, and there's a lot of strategic political questions about whether you can garner majorities for something this big and broad," he said.

But many climate-change policies elsewhere in the world are ambitious and have multiple goals, according to David M. Driesen, professor at Syracuse University's College of Law, who focuses on environmental law, law and economics, and constitutional law. "I see its primary potential as being a ... populist political proposal that might not pass now, but if it's done right and messaged right, might have the capacity to help change electoral results," he said. "And that's what's needed. We're going nowhere unless there are bold proposals put forward on messaging the shifts where the polity is at on these things."

Orts, Spence, and Driesen shared their thoughts on what the Green New Deal could achieve on the Knowledge at Wharton show on Sirius XM. (Listen to the podcast at the top of this page.)

According to Orts, the central takeaway from the Green New Deal is its emphasis on moving away from fossil fuels to renewables to combat climate change. "There does need to be a major investment and shifting in where we're sourcing our energy," he said. "You need to move to a more electrified economy, and you need to move away from fossil fuels, particularly coal and oil." He said the proposals drive home the point that "you need a major effort that can only be directed by a national government policy."

"If it's done right and messaged right, it might have the capacity to help change electoral results."

—David M. Driesen

Implementation Challenges

Incentives in the right places would be needed to implement specific proposals, said Orts. "You have to give co-benefits to everybody else," he explained. "You can't just put a charge on carbon, and everybody will say, 'OK, let's do it that way.' The problem is there's no political sell-ability for that. People see it as job-killing or business-killing." That issue can be addressed if "significant government investment" is combined with a commitment that it would bring "really good jobs — not just temporary jobs."

Spence also pointed to "tradeoffs in trying to accomplish two things at the same time," referring to the climate-change policies and green sector jobs the proposals seek. He acknowledged that renewable energy sectors such as solar promise more jobs, especially with a boom in the construction of solar rooftop panels, than what could come from the coal industry.

However, the Green New Deal talks about permanent union jobs, Spence pointed out. "There are more permanent union jobs in a coal plant or a nuclear power plant at the operational stage than there are in a wind or a solar or even a hydro station, modern versions of which are typically operated remotely from the control room with nobody on site. So, we have to think through these various tradeoffs. Some really tough decisions have to be made when we translate these goals that are in the current resolution into actual policy."

"There are a lot of questions about how you get from A to Z in each of these questions, and there's a lot of strategic political questions about whether you can garner majorities for something this big and broad."

—David Spence

For instance, Spence noted that shifting the production of solar panels from China to the U.S., for example, would mean "more expensive solar panels." While he did not rule out the possibility of those tradeoffs being resolved, he said they are "papered over right now in this resolution."

Driesen agreed that those tradeoffs are challenges. "They have to figure out how to take care of workers in certain regions that produce oil or mine coal," he said. "Otherwise, they're going to be vociferous opponents, no matter what good things happen in the rest of the economy."

In addition to promising permanent, new jobs, the "messaging" around the proposals could point to the "indirect job benefits" that could result from a shift away from fossil fuels to renewable energy, said Driesen.

Where Are the New Green Jobs?

Significant opportunities exist to create those green jobs, Spence said. "We have more than 10 gigawatts of solar in the queue waiting for permission to get built [in Texas]." However, it would call for "a heavy-handed effort and top-down leadership," he said. "Building the kind of grid that would connect all the windy and sunny areas to the places where the electricity is needed will require a lot of transmission lines. [But] it's hard to build transmission lines because of political and local opposition and some impediments built into the way we site transmission lines legally — changing all of that is not popular. We would have to essentially ram it down the throats of states and local governments to get that done in the timeframe we're talking about."

Driesen could try to stimulate renewable energy production in Texas, which has the Permian Basin, the top oil-producing region in the country. But that challenge could prove huge, monumental issues. At last count, Texas employed 352,371 people in the oil and gas industry, accounting for 40% of all U.S. oil and gas jobs, according to the 2019 State of Energy Report by the Texas Independent Producers and Royalty Owners Association, released a week ago.

Driesen also found flaws in the Green New Deal's proposal to retrofit all buildings in the country to make them energy-efficient and create new jobs. "[Even] in countries or states with ambitious energy efficiency policies for buildings, they are usually focused on new construction," he said. "They barely touch even big renovations, but it seldom gets [implemented] on existing buildings." However, even as the proposal is "super ambitious," serious-minded policymakers could find the financing for it and implement it. It would bring "a ton of skilled employment," he added.

Spence was "skeptical" about whether the shift to renewable energy sources could be achieved in the 10-year timeframe set out in the proposals. The decarbonization process will have to proceed sector by sector, he said. While the route to achieving that in the electricity generation industry is "fairly clearly" visible, "we are on the learning curve" in other sectors such as transportation, manufacturing, and agriculture, he added. He pointed out, for example, that "electric vehicles are a minuscule percentage of our transportation fleet, and the oil sector is [currently] serving that market."

It's a case of no pain, no gain, as Orts saw the implications of moving away from fossil fuels. "If you're going to make some serious change that moves the needle on climate change, you need to make these big changes," he said. "That is going to be disruptive to the [oil and gas sector] jobs in Texas [and elsewhere]. It is going to be a job hit."

Too Little Time, but Worth the Effort

According to Orts, the 10-year timeframe proposed in the Green New Deal to go carbon-neutral is also worrisome. He noted that climate scientists have set an aggressive target of 12 years to achieve that. A United Nations report last October said global net human-caused emissions of carbon dioxide would need to fall by about 45% from 2010 levels by 2030 and reach "net zero" around 2050. "This goal is achievable, but it will require enormous commitment by governments, businesses, and nonprofit organizations to mobilize support," Orts told Knowledge at Wharton in October.

Realistically, "it's not going to happen within a 10-year frame," Orts said. "But that doesn't mean it's not a good idea to get behind this program." He noted that the Clean Air Act of 1970 and the Clean Water Act of 1972 were greeted with skepticism, but "you have had progress in the decades since." Much could be achieved in tackling climate change if it were to be treated like a national emergency, he said, but he clarified that he wasn't precisely advocating that. "You need to try to move the ball forward, even though there's going to be a lot of difficulty in doing that [with the latest proposals]."

"One of the tensions we're not facing is there are a lot of people who think that 100% renewable is impossible, period," said Driesen. "The people who think it's possible are relying on technological advances and battery storage that haven't happened yet. It seems to me that if you need a shorter timeframe, which strengthens the case for nuclear [power]. And then there is the price tag," or challenges in financing those shifts.

Political Challenges

The resolution has 68 co-sponsors in the House and 11 co-sponsors in the Senate. Its support base includes 2020 presidential candidates such as Kamala Harris, Elizabeth Warren, Cory Booker, and Kirsten Gillibrand, even as some sounded cautious, like Amy Klobuchar, the Democratic senator from Minnesota..

"You can't just put a charge carbon, and everybody will say, 'OK, let's do it that way.' The problem is there's no political sell-ability for that. People see it as job-killing or business-killing."

—Eric Ort

Republicans have jeered and mocked the New Green Deal. President tweeted that it would "permanently eliminate all planes, cars, cows, oil, gas & the Military, even if no other country would do the same." Wyoming Senator John Barrasso, chairman of the Senate's environment committee, called it "a socialist manifesto that lays out a laundry list of government giveaways, including guaranteed food, housing, college, and economic security even for those who refuse to work."

Despite the plan's weak spots, Orts described McConnell's eagerness to put the proposals to the vote as "a miscalculation" — that it would end up being "embarrassing" for the Republicans. "I would suggest to the Democrats to call their bluff and vote this in," he said.

In trying to address climate change, the broad sweep of the proposals could end up being counterproductive, according to Spence. "When you add additional dimensions to a piece of legislation, it might help build a broader coalition or cause people to peel off."

Driesen predicted little or no chance for a majority in Congress to vote in the Green New Deal but reiterated that its biggest virtue lies in the ability to spread the word on the urgency for serious action on climate change. "The Republicans in the Senate [where they have the majority] will not permit it," he said. "So, the goal of this has to be to have a messaging strategy or a coalition or both that will change the electoral dynamic and, therefore, the breadth of this."

According to Orts, Republicans should apply a larger lens to the Green New Deal rather than viewing it as an opportunity to score political brownie points. "It's not enough to just come in and say, 'Let's embarrass the Democrats' by bringing this to the floor and calling it socialism or something like that," he said. "There has to be something that [the Republicans are] bringing."

Orts also decried the tendency to label the proposals as socialist or neo-liberal simply. Instead, he wanted the conversation to address questions such as: "How do we have a smart grid? How do we make a transition? What policies can we put in place? What makes sense economically, what makes sense for business, and what makes sense for people? Is this going to hurt jobs?"

Author's notes:

After posting notes at the beginning of this article, I reviewed it for applicable content and was pleasantly surprised to find something I would like to dwell upon at the end. It's the last two paragraphs, and I found them to be serious words we all should take heed of and encourage our politicians to notice.

- This subject matter is not a bush-league issue on which we can brush off and/or score political points. This is our planet; it's time all of us put on our big Boy and Girl clothes and constructively discuss a mutual Blue Ribbon August panel of experts selected for their constructive and problem-solving abilities without dyed-in-the-wool political ties.

- I foresee no time limit; it will take whatever it needs to develop reasoned solutions or constructive methods to address the first things we can change and apply solutions or potential fixes and move on to the next critical path issue. Our home, the Earth, is in trouble, and I believe the prognosis is not good. Yes, it is projected to get worse, and the huge population of countries needs a wake-up call.

- Our people are a concern, and these gentlemen aptly address the issue, such as the policies we can put in place. What makes sense economically, what makes sense for business, and what makes sense for people? Is this going to hurt jobs?"

I am totally impressed with these gentlemen and the schools they represent. I look forward to hearing more from them.

Biden's Green New Deal: A Recipe for Failure

Joseph Toomey

Independent Management Consultant

February 27, 2021

Author's notes:

I have chosen this PART 1 article, even though it is dated, for its forthright directness and candid remarks defining issues with politics, the NGD, jobs, energy issues, economics, issues with electric vehicles, productivity, and many related items. You deserve to be exposed to a fresh and reasoned presentation of facts, which most articles do not present well. Enjoy his work and others I may add to this book's content.

Thank you, Mr. Toomey, for using your unabridged article. You have my earnest respect. I have tried to contact you for a personal telecom meeting through several sources and have not been successful. Your materials are what this book needs for its readership, and I am going to use them, hopefully, with your freely given permission. Thank you for the superior article.

This is Part 1 of the analysis of the impact that Biden's Green New Deal is likely to have on U.S. economic performance, job creation, and energy security. Biden's energy and climate plans are just a reprise of Obama's green energy failure but on steroids.

- Biden's focus on phasing out fossil fuel has already destroyed thousands of high-paying union jobs without any replacement job creation anywhere.
- Green energy is a very low-productivity energy source that saps the main engine of economic growth.
- Obama's green energy program helped drive down productivity to one of the lowest levels in the post-War era.
- Massive switchover to EVs will merely displace conventional auto sales, offering no net economic advantage.

- EVs are likely to involve higher levels of foreign content, worsening trade and current account deficits.

- Mandated investments in EV charging networks will siphon resources to create a very low-productivity infrastructure that is less efficient, more resource-intensive, and more time-consuming.

- Biden's Green New Deal is likely to re-launch the ridiculous green jobs counting frenzy that became a national pastime during the Obama years.

- Existing green energy jobs by green activist groups bear little or no relationship to BLS job counts or to the value of green energy output.

- Worker productivity degradation is exemplified by solar energy, which requires 30 times more, and wind energy, which requires 5 times more workers per unit of energy produced than natural gas.

- Green job creation comes at staggering price tags, often tens of millions of dollars in government support per full-time job created.

- Unionization rates among green jobs occupational classifications are among the lowest of industrial workers in the U.S.

- The likelihood that Biden can deliver on his promise to create millions of unionized green energy jobs is incredibly low.

- Deficit spending to jump-start a green energy program would be economically depressing in a variety of ways that include crowding out private sector investment, increasing the sovereign debt load, undermining the value of the U.S. dollar, boosting interest rates, spurring inflation, reducing productivity, worsening the balance of trade and current account deficits, driving energy prices higher, making industry less competitive, curtailing economic growth, lowering personal incomes, and eroding the country's energy security

Introduction

Biden campaigned on a promise to unleash a tsunami of debt-fueled deficit spending whose total price tag would amount to $5 trillion to implement his Green New Deal. As alert observers well know, his program didn't involve a single syllable of originality. It was a word-for-word reprise of the failed Obama green energy-climate program distinguished only by its order-of-magnitude increase in spending scope and level of hysterical amplitude. This essay will examine economic, workforce, and energy output data to show how Biden's program is indistinguishable from Obama's, why Obama's program was such a colossal failure, and why Biden's program will also fail even more catastrophically.

Biden's inauguration address was where he spoke about his unique hearing ability. "A cry for survival comes" from the planet itself. A cry that can't be any more desperate or clear," he proclaimed. Biden assured the world it was "a climate in crisis." Never mind that this "cry" is one that, like the voices in their heads telling them the 2016 election was stolen. Still, the 2020 election was not, it can only be heard by a select few people.

Biden spoke of a "battle to save our planet by getting climate under control battle to save our planet by getting climate under control" in his victory speech immediately after the election. Noticeably absent in those two seminal speeches was any concrete plan to create high-wage jobs, grow personal and household incomes, or boost the output of farms, factories, and businesses. all the signs pointed to a high likelihood of exactly the opposite outcome occurring. And never mind that to sell his plan, he'd need to assure Americans that a devastating freeze in Texas that knocked out windmills and forced power and water offline for days on end was yet another proof point of global warming, a painful reminder that America needed his Green New Deal to forestall the runaway atmospheric heating that was causing people to freeze to death in their homes.

Everything about economic and energy policy from the prior administration needed to change. That was the administration that had, for the first time since 1957, empowered the U.S. to generate more primary energy than the nation had consumed, that had permitted the country to become petroleum self-sufficient on a volumetric basis for the first time since prior to World War II, that had increased the monthly pace of full time job creation among women during the pre-pandemic period by 70% compared to the last 4 years of the Obama administration, that had boosted the rate of average annual payroll wage growth by 44% over that same time frame, that had posted the lowest rate of unemployment in more than 50 years, that had boosted the annual pace of industrial production by a factor of 13 (i.e. 1,302%) from the prior 4 years, that increased average monthly personal income growth by 80% and had doubled monthly per capita personal income growth, that restored a robust 4.1% annual rate of new factory order growth after a 4 year period of decline, that had seen the monthly rate of high-wage goods-producing (mining, construction, manufacturing) job creation increase by 35.2%, that had witnessed a 25% reduction in food stamp enrollment and a 27% reduction in benefits payout from the Obama caseload still lingering 4 years after the recession had ended, etc.

The Biden agenda was clear: Everything had to change. Judging by Biden's rhetoric thus far, it seems highly likely that everything will change.

Biden Central Planning

The newly inaugurated president wasted no time reversing his predecessor's job-creating energy policies. Within hours of being sworn in, Biden signed an executive order revoking the Keystone XL pipeline permit. That move resulted in the immediate loss of thousands of high-wage unionized construction and engineering jobs. The Keystone XL pipeline had become a curious touchstone among environmental activists, enduring an undeservedly long and torturous history before Biden's apparent *coup de grâce*.

Barack Obama was one of those activists. Despite the State Department issuing the first of its five final environmental impact statements in August 2011 granting approval to the proposal, the Obama administration repeatedly interfered to slow-walk it in a "defeat by delay" process. Obama would later issue a veto in hopes of killing the project, one of only 12 vetoes he'd issue during eight years in office after Congress voted overwhelmingly to approve the project in the Keystone XL Pipeline Approval Act.

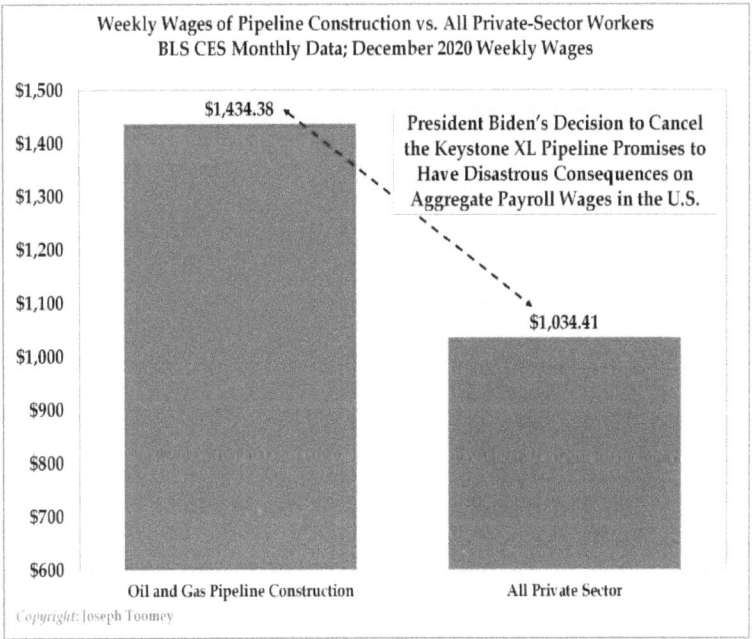

Weekly Wages of Pipeline Construction vs. All Private-Sector Workers
BLS CES Monthly Data; December 2020 Weekly Wages

$1,434.38

President Biden's Decision to Cancel the Keystone XL Pipeline Promises to Have Disastrous Consequences on Aggregate Payroll Wages in the U.S.

$1,034.41

Oil and Gas Pipeline Construction   All Private Sector

Copyright: Joseph Toomey

SOURCE: BLS Monthly Current Employment Statistics

Two months after Trump took office, he issued a permit by executive order after the State Department granted the last of its final approvals. A State Department study estimated the total impact of the KXL Pipeline from construction, extended supply chain, and indirect or induced employment totaled 26,100 jobs. As union criticism mounts, thousands of union workers are thrown out of work. Biden's press spokesperson is heard offering only snarky answers too seriously-posed questions about where laid-off workers can go today to find green jobs, and the administration begins its 10 million clean energy jobs tidal wave with a vast job deficit, serious union opposition, and some really bad political optics.

In addition to the pipeline permit revocation, Biden announced an immediate moratorium on drilling permits on federal lands. Defenders of the measure hastened to point out that a moratorium was not an outright ban. The measure only covered a mere 90 days. But observers immediately recognized this as a rerun of Obama's 180-day moratorium on outer continental shelf drilling after the Deepwater Horizon disaster. Obama would later lift that moratorium. But, in its place, he would establish a set of onerous new regulations that functioned as a de facto extension of the moratorium. Five months after the administration announced its lifting of the moratorium, the Interior Department had still not managed to approve a single new exploratory deep-water drilling plan.

Biden had solemnly promised on multiple occasions during his campaign that he was "not banning fracking in Pennsylvania or anywhere else." What could be more unequivocal? "I will not ban fracking. Period," he said to a Dallas, Pennsylvania audience, adding, "I will protect jobs in Pennsylvania." So naturally, when it came time to protect jobs in Pennsylvania, Working-Class Joe abstained. When the Delaware River Basin Commission, ignoring the incontestable findings of the Obama EPA, voted to ban fracking, Biden punted. So yes, he didn't precisely "ban" fracking and thereby destroy Pennsylvania jobs. He allowed others to do it for him.

It was familiar with Obama's style. Offer an endless series of meaningless platitudes about the need for new sources of energy or the importance of protecting jobs, but then do exactly the opposite. Despite his rhetoric, in the wake of the Deepwater Horizon spill, Obama would cancel four lease sales in Alaska, ban both shallow and deep water drilling in the Gulf, lift the Gulf drilling ban but later impose a de facto moratorium by spiking the approval process, and open the Eastern Gulf, Atlantic, and Pacific coastal areas to drilling only to close them off again a few months later.

Close observers had also seen a similar pattern in Obama EPA rulemaking. In April 2010, Obama's EPA issued "guidance" on a controversial coal mining technique called mountain-top removal. That was guidance, just a "recommendation." But the agency then used that same "guidance" as its authority to shut down a coal mining operation in Logan County, West Virginia, a regulatory action that put 250 employees out of work. With nearly 50 years of high-level insider experience in government, Biden needed no education in political cynicism that his neophyte predecessor could never hope to master. Thus, a "temporary" moratorium on drilling provides absolutely no reassurance to those whose jobs are dependent on a robust "all of the above" energy policy. As we learned from bitter experiences during the Obama years, all of the above means nothing below the surface.

John Kerry, Biden's Special Presidential Envoy for Climate, claims laid-off pipeline workers can "go to work making the solar panels." The obvious trouble is those jobs Biden has killed are gone today, while the ones he claims he'll create are far off in the future, or may never materialize. Even if they do appear — and there's absolutely no certainty they will — they'll pay far less and likely occur in locales far from the high-paying union jobs he has now eliminated. Heather Richards of *Energy & Environment News* reports that:

"A union jobs retraining program for miners run since the 1990s in Pennsylvania and West Virginia has supported thousands of workers. But the average wage they earn after that program is about $14 an hour.

Those were workers who had been making as much as $82,000 per year, or $42.71 per hour for a 1,920-hour work year. But don't let that bother you. *NPR* agrees workers will get far less pay, but they'll be rewarded with more passion. they said that. That will sure come in handy when it comes time to pay mortgage bills and put kids through college.

On his first day in office, Biden announced the U.S. would rejoin the Paris Climate Accord. This treaty obligates some but by no means all signatory nations to reduce carbon dioxide emissions. Biden's appointment of John Kerry, an aloof, condescending, private jet-traveling career politician, signals that he would effectively become the nation's real Energy & Environment Secretary. The signs are not encouraging. Biden appears more intent on pleasing Fabian socialists of the European Union at the expense of American workers if Biden environment advisor Gina McCarthy's rhetoric is to be believed.

And why shouldn't it? It's a certainty U.S. emissions cuts won't be matched by the world's largest emitter, China, or by India, the world's third largest and most rapidly growing source. To honor these draconian commitments, deep cuts in the industrial workforce that will lead to a permanent loss of tens or even hundreds of thousands of high-wage, unionized, goods-producing jobs will need to be made.

A few days after the inauguration, Biden promised to put the immense purchasing power of the federal government behind a plan to acquire a fleet of non-emitting electric vehicles, claiming the measure would lead to significant job gains. Needless to say, one need never be burdened with explaining why acquiring non-emitting cars will create loads of jobs while existing federal fleet vehicle purchases of conventional vehicles presumably would not. And never mind that the federal fleet averages about 6,000 miles per year. That is less than half the 13,476 annual mileage per vehicle the average private vehicle achieves. An NBER study of privately-owned EVs published in 2021 found the average annual mileage per vehicle was just 5,300 miles. Thus, the far higher acquisition price tags of EVs will have fewer miles to monetize their lower ongoing operating cost, making them far less economical than a conventional model choice would be for a typical motorist.

We also learned that Biden is planning to force American taxpayers to fund a nationwide electric vehicle (EV) charging network to induce consumer EV adoption. Some 500,000 EV chargers will be built and deployed by 2030. Like Soviet-era central planners, Biden has decided he can outthink consumer tastes and trends. We can hardly wait until he decides how many brands of washing machines or toothpaste consumers require. None of this bodes well for the American economy.

We'll examine in detail how these programs have worked in the past and what impact they will likely have on jobs, productivity, economic growth, balance of trade, consumer prices, and a host of other considerations. It already shows issues with the cars burning, inoperative on very wet roads, wintry weather and the lack of charging stations. We will monitor industry and the impacts on fuel-driven vehicles. As of December 20, 2024, there are 140,000 fast and slow chargers.

To support them, the U.S. needs 28 million more EV charging ports. Many of those can be private since most EV drivers charge at home. 2023 reports indicate approximately 16,000 at 38,000 locations exist. But we'd still need more than 1.2 million public ports, or about 1 million more than we have now. At the current rate of 1,000 ports a week, which would take 19 years, according to a Dec 20, 2024, report.

All hail to the New Green Deal forced electric vehicles mandate by 2030. Just think about whether we have buses, trucks, airplanes, and even more EVs coming on stream if they want.

Labor Productivity and Economic Growth

As noted above, Biden's plan to revive the failed Obama green energy plan with an order of magnitude larger infusion of debt-fueled spending bodes poorly for long-term economic growth prospects. Forget the rhetoric about reviving the economy or empty promises to "Build Back Better" by employing borrowed money to incentivize costly energy sources that rely upon foreign-sourced components, energy output that must be heavily subsidized to be fed into highly regulated and distorted markets just to allow them to succeed. The plan is doomed.

Two basic factors contribute to long-run economic growth. They are:

(1) increases in labor inputs from an increase in hours worked by an expanding labor pool or

(2) technology advances that improve productivity (an increase in value produced per hour worked).

That's it. There is nothing else on offer like **SOURCES** on economic growth. All the other factors that we typically examine as components of economic output, like personal consumption expenditures, business investments, trade surpluses, government expenditures, and so forth, are actually **USES** of growth.

Most economic growth comes from productivity improvements. As Chad Stone of the Center on Budget and Policy Priorities testified in April 2017:

> "*Broadly speaking, there are two main sources of economic growth: growth in the size of the workforce and growth in the productivity (output per hour worked) of that workforce. Either it can increase the overall size of the economy, but **only strong productivity growth can increase GDP and income per capita**. Productivity growth allows people to achieve a higher material standard of living without having to work more hours or to enjoy the same material standard of living while spending fewer hours in the paid labor force.*"

Within a month after inauguration day in 2009, Congress passed and Obama signed into law the American Recovery and Reinvestment Act (ARRA), unleashing an estimated $831 billion flood of deficit spending into a still-contracting economy. A notable feature of the 2007-to-2009 recession that the bill attempted to alleviate was its deep impact on job counts and hours worked compared to the deep recession in the early 1980s. Payroll employment had contracted 6.7% by February 2010 from its prior peak in January 2008. Meanwhile, during that same time, average weekly hours worked plunged from 34.4 to 33.8 due to a large loss of work When these two factors are combined, economy-wide employee hours worked declined by 8.6% on an annualized basis.

By contrast, the deepest job count decline during the Carter-Reagan recession was far milder. Parenthetically, the National Bureau of Economic Research counts 2 recessions over the 1980-to-1982 time period, but it's entirely reasonable to view it as a single, torturous, long-duration contraction interrupted by a short 2-

140

quarter growth spurt. The worst recorded jobs decline was only 2.4% even though the 1980s recession had seen the U3 unemployment rate soar to 10.85% and stay in double-digit territory for an agonizing 10 months. That compares to the Bush-Obama recession, where U3 unemployment hit just 10% — actually 9.982833% if anyone's keeping score — and did so for only one single month.

Green Energy: A Recipe for Productivity Erosion

Why is this important and how does it relate to likely economic outcomes from the proposed Biden green jobs program? Remember that productivity improvement is the most important contributor to long-run economic growth. This is chiefly due to the diminishing returns from labor over the long run. For instance, a skilled bricklayer today can build a brick walkway of a given length and width at about the same time as it took his grandfather. Due to the nature of their work, most labor-intensive service economy occupations can achieve little improvement in marginal return from the effects of productivity-enhancing innovation.

This is in stark contrast to a goods-producing industry like automobile assembly, which has seen significant and continuous labor productivity gains from the employment of capital goods like robotic assembly tools, material substitutes, and other automation features. Indeed, the gradual shift of the U.S. from a goods-producing to a services-based economy, where a large portion of economic output is now derived from slow productivity-growing personal service labor, is the primary reason why U.S. economic growth rates today are substantially lower than decades earlier.

When U.S. non-farm labor productivity over the past 17 years is examined, one notices a pronounced secular downtrend beginning in 2004. But suddenly, in 2010, a sudden increase in *measured* productivity occurred in the quarters immediately after ARRA spending began flowing through the economy. Enormous stimulative fiscal impact (numerator) unleashed on an economy where labor hours were rapidly contracting (denominator) sent *"measured"* labor productivity soaring, albeit momentarily. It is as the chart below clearly depicts.

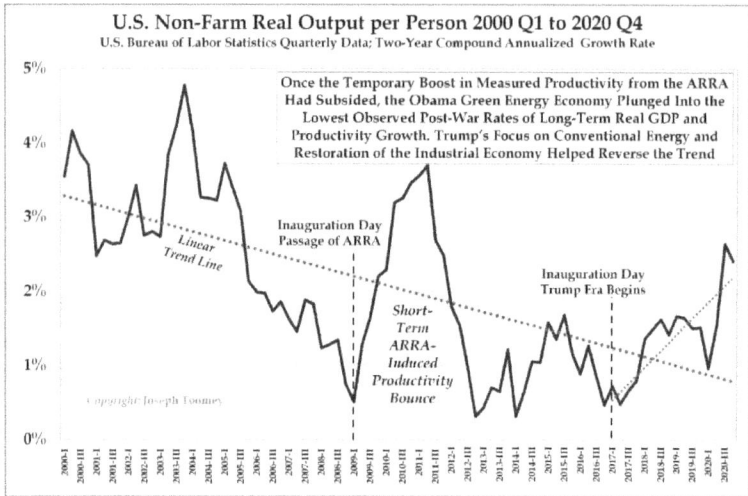

U.S. Non-Farm Real Output per Person 2000 Q1 to 2020 Q4
U.S. Bureau of Labor Statistics Quarterly Data; Two-Year Compound Annualized Growth Rate

Once the Temporary Boost in Measured Productivity from the ARRA Had Subsided, the Obama Green Energy Economy Plunged Into the Lowest Observed Post-War Rates of Long-Term Real GDP and Productivity Growth. Trump's Focus on Conventional Energy and Restoration of the Industrial Economy Helped Reverse the Trend

Linear Trend Line

Inauguration Day Passage of ARRA

Inauguration Day Trump Era Begins

Short-Term ARRA-Induced Productivity Bounce

Copyright Joseph Toomey

Once the economic stupor of massive deficit spending had subsided, the low rate of Obama's economic growth and low productivity began to assert itself. Real GDP and real output per hour worked (except for the Carter runaway inflation era) posted the lowest sustained levels ever recorded in the U.S. post-war period as the nearby chart shows. The low productivity engineered by Obama policies was clearly aided and abetted by increasing deployment of productivity-destroying green energy investments like solar, wind, biofuels, and electrified transport. During the Obama years, these green energy pursuits were conducted at a relatively small scale compared to Biden's announced ambitions. For instance, even after all of Obama's Herculean exertions, wind and solar output accounted for just 10.4% of U.S. electricity generation in 2020 and 4.6% of U.S. primary energy. Biden seeks to convert this wind and solar penetration in electricity generation to 100% by 2035, a nearly 10-fold increase.

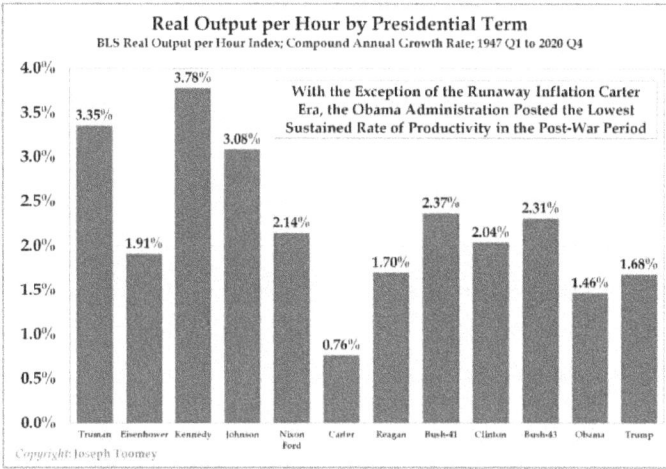

Real Output per Hour by Presidential Term
BLS Real Output per Hour Index; Compound Annual Growth Rate; 1947 Q1 to 2020 Q4

With the Exception of the Runaway Inflation Carter Era, the Obama Administration Posted the Lowest Sustained Rate of Productivity in the Post-War Period

Truman 3.35%, Eisenhower 1.91%, Kennedy 3.78%, Johnson 3.08%, Nixon Ford 2.14%, Carter 0.76%, Reagan 1.70%, Bush-41 2.37%, Clinton 2.04%, Bush-43 2.31%, Obama 1.46%, Trump 1.68%

Copyright: Joseph Toomey

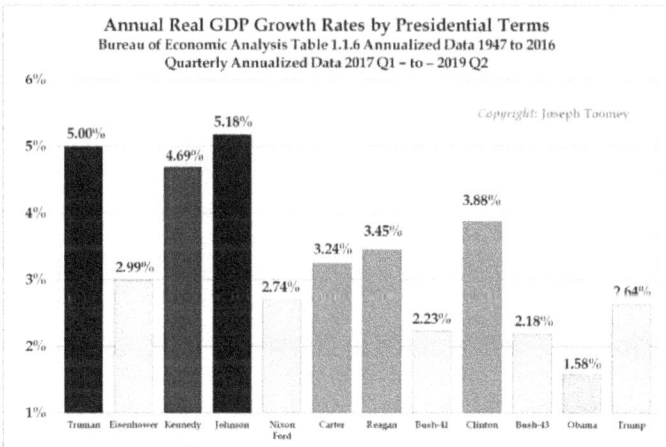

Annual Real GDP Growth Rates by Presidential Terms
Bureau of Economic Analysis Table 1.1.6 Annualized Data 1947 to 2016
Quarterly Annualized Data 2017 Q1 – to – 2019 Q2

Copyright: Joseph Toomey

Truman 5.00%, Eisenhower 2.99%, Kennedy 4.69%, Johnson 5.18%, Nixon Ford 2.74%, Carter 3.24%, Reagan 3.45%, Bush-41 2.23%, Clinton 3.88%, Bush-43 2.18%, Obama 1.58%, Trump 2.64%

Electric Vehicles and Economic Growth

While much of the attention thus far has been focused on solar and wind power deployment, there is another critical productivity-degrading feature of the Obama-Biden green energy program. Obama campaigned on a promise to put 1 million electric vehicles on the road by 2015. At least, in February 2007, when Obama threw his hat into the ring, the U.S. only derived about 62.5% of its primary energy and a paltry 49.2% of its petroleum consumption requirements from domestic supply sources. So, there was a putative import-displacing rationale to his proposal.

143

Nevertheless, the plan was a ridiculously delusional overreach. Thankfully, for the sake of job-seekers and the economy, it didn't come close to fruition. Needless to say, Biden plans to triple down on Obama's program by using government resources to promote or likely emulate California governor Newsom's plan to force electric vehicle (EV) adoption.

The first widespread fallacy that green energy boosters make is their contention that EV adoption will result in a significant economic boom because new cars will be needed. That fallacy is easily debunked. Sure, auto assembly plants will be humming with activity, churning out the vehicles of the future. Look at all the jobs and income that will be created. Assembly and sub-assembly parts suppliers will be kept busy manufacturing components that are used in EVs. Truckers will receive a steady flow of orders to haul parts in and finished vehicles out of vehicle assembly plants or raw materials in and finished assembly parts out of parts suppliers' plants. And so on.

But that same activity is already happening today. Is there any reason to believe that a switchover to EV adoption will suddenly translate into far larger unit sales of finished autos and trucks? If so, it is a well-kept secret, as the chart of unit sales of autos and light trucks over the last 3 decades demonstrates.

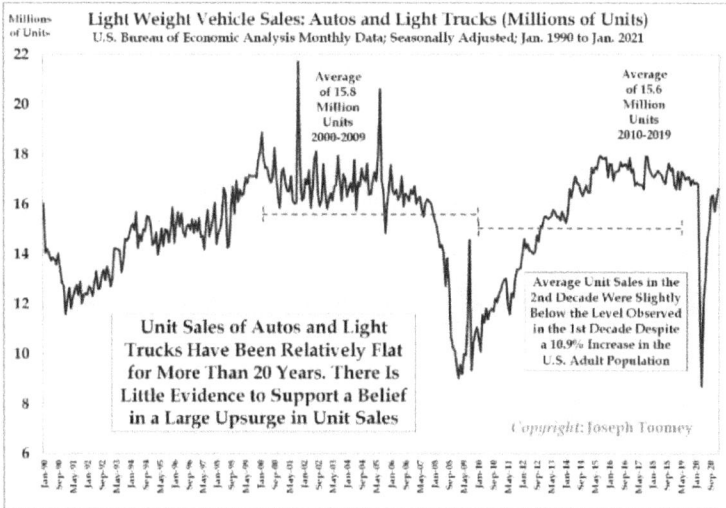

Light Weight Vehicle Sales: Autos and Light Trucks (Millions of Units)
U.S. Bureau of Economic Analysis Monthly Data; Seasonally Adjusted; Jan. 1990 to Jan. 2021

Average of 15.8 Million Units 2000-2009

Average of 15.6 Million Units 2010-2019

Average Unit Sales in the 2nd Decade Were Slightly Below the Level Observed in the 1st Decade Despite a 10.9% Increase in the U.S. Adult Population

Unit Sales of Autos and Light Trucks Have Been Relatively Flat for More Than 20 Years. There Is Little Evidence to Support a Belief in a Large Upsurge in Unit Sales

Copyright: Joseph Toomey

In addition, there is the critical aspect of domestic content of existing conventional vs. EV cars of the future. Unless major EV components like batteries and electric motors are fabricated and assembled in the U.S. — and there is little evidence suggesting EV domestic content percentages will materially improve — there is no net benefit compared with conventional automobiles and trucks. Indeed, China controls 70% of global EV battery production.

The U.S. controls less than 10%. All of the nickel, cobalt, and lithium used to produce EV batteries is sourced from outside the U.S. While the net foreign vs. domestic content calculations are complex, the prospect for vast net economic benefit from domestic EV production is not looking too rosy at first glance.

So, how could EV adoption lead to substantial economic degradation through productivity erosion? What is the basis for economic pessimism over the Biden EV plan, especially after its boosters claim it will be so economically beneficial? The primary reasons for relatively slow EV adoption are higher EV prices relative to comparable conventional models, consumer hesitancy over EV range anxiety, and the sparse availability of EV charging infrastructure.

Price issues are being addressed through massive federal and state incentives and continuous reductions in lithium-ion battery costs. Range anxiety is being addressed through improvements in battery technology, larger capacities, and range-extending devices. That leaves charging infrastructure as the last consumer hurdle to overcome. And therein lies the rub, economically speaking.

The Broken Windows of Electric Vehicles

French economist Frédéric Bastiat published a pamphlet in 1850 whose title is translated as *"What Is Seen and What Is Not Seen."* Many people, even including Nobel laureate Paul Krugman, seem to accept on faith the delusion that "war is good for the economy." Bastiat showed through his "parable of the broken window" — more commonly referred to as "the broken window fallacy" — that such is not the case. Simply stated, as a way to boost economic activity, one might go around the neighborhood breaking windows.

Doing so would necessitate copious economic activity among glaziers who perform window repair. They, in turn, would keep the foundries busy making replacement windows. Those foundries would place orders with raw material suppliers and pay wages to workers to produce new windows. Haulers would be kept busy delivering raw materials and finished goods.

And so, on goes the myth. Yet, in the end, once all the windows were fixed, we would have seen a massive quantity of scarce economic resources expended just to return us to our original state of social welfare. These resources could have been used to create incremental welfare improvements that must now be foregone. Like war, broken windows are merely an opportunity cost, and thinking otherwise is charlatanism. Prestigious publications like *The Economist* and presidential economic advisors like Larry Summers or Nobel laureates like Paul Krugman regularly demonstrate they are by no means immune to the fallacy.

To overcome the last EV consumer adoption hurdle (i.e., range anxiety), federal and state governments will likely need to both mandate and incentivize a massive and breathtakingly expensive

146

charging network. In many states, buildings, workplaces, roadways, and retail centers of a certain size will likely be required to install an ever-increasing number of charge points throughout their parking lots and garages. Public streets and roadways will also be torn up to install charging stations and load-carrying infrastructure needed to manage higher loads.

The political effect will be to help overcome consumer anxiety about the availability of vehicle charging. But the economic effect will be to degrade GDP growth as massive expenditures that could otherwise be used to finance improvements in societal well-being will be siphoned off to create a vehicle refueling infrastructure that is less efficient, more resource-intensive, and more time-consuming than the one that already exists. The massive flood of expenditure will be merely used to create a network that adds nothing to net societal welfare and will likely subtract from it. The charging network is an example of the broken windows fallacy played out on a massive, nationwide scale.

Furthermore, each charge point is likely to suffer significantly lower utilization than existing fuel pumps. This author's nearby suburban fueling station is a good example of that shift. Throughout daylight hours, each fuel pump is occupied, on average, about half of the time with motorists refueling vehicles. The average dwell time per refueling event is 6 to 7 minutes, but to illustrate the point, assume it's 10 minutes.

During that time, the average fuel expenditure at today's prices is between $30 and $45 which supplies enough stored chemical energy to provide a motorist with 275-to-350 miles before refueling. Thus, each fueling pump delivers an average of 7-to-10 days of motoring (based upon an average annual vehicle distance of 13,476 miles) within 10 minutes before it is made free to accept another customer. Each pump can accommodate perhaps 30 refueling events in a typical 10-hour day (although the station is open 24 hours and many motorists refuel during nighttime hours, improving utilization and equipment productivity).

With an EV charge point, a motorist will pull into a charge point-equipped parking spot and occupy it for an exceptionally long duration, maybe even all day. There are two reasons for that. It takes significantly longer to recharge an EV and because the primary purpose of that parking spot is vehicle parking. During that dwell time, the vehicle might accept an average of 30-to-45 kilowatt-hours of electrical energy which U.S. residential consumers an average of $0.1323 per kWh in 2020. But commercial charging prices are 3-to-4 times that amount. That will provide about 90-to-135 miles of driving distance at 3 miles per kWh, although lithium-ion batteries can only be charged to 80% of their capacity since charging above this amount will result in battery damage.

The cost of each charge point installation may be maybe $50,000 or even more, given that nearly all will require retrofitting existing streets, lots, and garages. New buildings may entail a lower cost than this on average. But it will probably take a century to completely replace the existing stock of buildings. Each charge point will likely be occupied only once during an entire calendar day — or each of 5 days per week for workplace infrastructure — delivering about $4-to-$6 of revenue potential per day. The daily profit potential will be far less since charging resellers will need to pay a bulk commercial rate, which was $0.1066 per kWh in 2020, to acquire the electricity from utility generators.

Commercial charge distributor Electrify America has a solution to this economic productivity degradation problem. It has established a pricing structure that essentially requires EV motorists to become captives during charging. Electrify America allows a "10-minute grace period" after charging is complete.

Dwell times exceeding that are charged an occupancy penalty of $0.40 per minute. If a motorist is delayed by an hour, the penalty would amount to $20.00. The electricity tariff price for Washington, DC motorists is $0.43 per kWh, which is 3.4 times the electricity price in 2020 for residents. If you live in a dwelling without home charging (a row home, an apartment complex, etc.) or simply don't want to bear the expense of rewiring your home to accept 240-volt charging, expect to waste 500-to-700 hours of time each year in EV recharging. Your car has become your new office or home.

Indeed, recognizing the entire economic case resides solely in asset utilization rather than in reselling electricity, commercial charge distributor Evgo doesn't even care how much electricity flow occurs. Motorists pay strictly according to occupancy time. In Washington, DC, the current tariff structure costs non-members $0.30 per minute on the fast Direct Current charger. The firm advises that a typical EV can achieve 90-to-120 miles in a 45-minute session, which would cost $13.50. If your vehicle has a 30-kW charge rate, the effective price is $0.60 per kilowatt-hour. Compare that to the average nationwide residential rate of $0.1323 per kWh. That's 4.5 times the nationwide rate. It amounts to between 11.25 and 15 cents per mile. An internal combustion vehicle that gets 25 miles per gallon refilling today at my nearby station would pay $2.39 per gallon for regular unleaded 87 octane gasoline, which costs 9.56¢ per mile.

On top of all that, most homeowners will likely choose to equip themselves with home charge systems to provide overnight charging. That may require a homeowner to replace the service entry cable to the home with expanded wattage capacity to accept a larger current load required for efficient recharging. That would entail thousands of dollars for equipment and retrofitting. The amount of poorly utilized, low productivity infrastructure investment required on average to support a single EV is simply staggering — and thus economically degrading. EV motorists will also find themselves charging every day of their lives compared to the conventional vehicle motorist visiting a refueling station once every 7-to-10 days. Thus, to avoid economic degradation, consumers will sacrifice loads of leisure time and surrender loads of cash. This damage will not appear in the National Income and Product Accounts maintained by GDP scorekeeper U.S. Bureau of Economic Analysis during the Biden term. But it will represent a significant decline in our standard of living.

To sum it up, EV purchases are unlikely to generate any net economic growth because they will merely be substitutes for conventional models. And anecdotal evidence thus far suggests these EV purchases may actually degrade economic performance with their higher price tags, which will reduce disposable income and a higher percentage of foreign content, which will worsen GDP growth as well as the balance of trade and current account deficits. Meanwhile, nationwide installation of, with its "broken windows" efficiency degradation, will entail ruinous and expensive tearing up of streets, buildings, parking lots, homes, and other places to install very low-productivity charging systems far less beneficial than existing refueling systems. The value of charging infrastructure investment required to support EVs will be many times the embedded investment required to support conventional vehicle refueling. None of these realities provides any reassurance that optimistic claims by Green New Deal proponents are well grounded.

The actual "net-zero" the U.S. is likely to achieve is net-zero economic growth as labor and capital productivity are deeply degraded, net-zero carbon dioxide reduction as production is merely shifted to other countries, and net-zero reduction in pollution as production and emissions are shifted to foreign countries with far less stringent environmental abatement restrictions.

The Green Jobs Promise

Biden's inauguration address included a strong admonition against dishonesty, particularly concerning public policy. "*We must reject a culture in which facts are manipulated and even manufactured,*" he claimed. Clearly, this doesn't apply to claims about the danger posed by human-caused climate change. It also does not apply to the actual cost of green energy. The net job-creating potential of his grandiose proposals is another area it does not cover. This includes the compensation levels of the green jobs he wants to create, as well as those he intends to destroy. The prospects for unionizing green jobs are also not addressed. Neither are the number of green jobs already in the economy. Lastly, it fails to cover the long-term economic impact his green jobs program is likely to have.

Barack Obama campaigned on a promise that if the U.S. *invested* $150 billion in green energy technology, it would "*create 5 million jobs that pay well and can't be outsourced.*" Needless to say, Obama didn't come close to fulfilling his job's promise. Rather than a flood of high-paying green jobs that couldn't be outsourced, taxpayers received a flood of red ink pouring out of an endless succession of crony-connected green energy megaflops.

ALL OF THE NEXT POINTS ARE NGD FAILURES OF THE OBAMA ADMINISTRATION

The Consumer Cost of the Green Finger on the Scale By shifting national incentives to disproportionately favor intermittent sources of power, such as wind and solar, the Biden Administration is artificially suppressing the actual cost of renewables through subsidies. Nuclear power generation continues to face lengthy licensing and permitting timelines leading to increased costs, limiting nuclear power's market potential as a low-carbon alternative.90 The Biden Administration's resistance to permitting and approving additional natural gas pipelines will be particularly harmful to areas of the country that rely on natural gas for heating or power generation. As Americans face increasing financial burdens imposed by the Biden Administration's policies, artificial government-mandated transformations of energy markets could not come at a worse time. The Subcommittee on Economic Growth, Energy Policy, and Regulatory Affairs held a hearing on March 29, 2023, featuring expert witnesses who testified about rising energy costs and their impact on inflation. One expert witness explained: Rising energy costs and inflation have created immense financial burdens on the American people. One in six American families is currently behind on electricity bills. 91 The cost for an average household has risen approximately $10,000 over the past two years. 92 Everyday goods like groceries and gas are exorbitantly expensive. […] These costs are squeezing the middle class and making it virtually impossible for low-income Americans to ever cross the middle-class threshold. Most concerning, some families have been forced to choose between powering their homes or putting food on the table.93 The long-lasting consequences of Biden Administration policies imposed on consumers are only just beginning to surface. America's electrical grid is under threat from shortsighted 89 The Power Struggle: Examining the Reliability and Security of America's Electrical Grid: Hearing Before the H. Comm. on Oversight and Accountability, Subcomm. on Economic Growth, Energy Policy, and Regulatory Affairs, 118th Cong. (Mar. 12, 2024) (statement of James P. Danly, former Commissioner, Federal Energy Regulatory Commission). 90 The Next Generation: Empowering American Nuclear Energy: Hearing Before the H. Comm. on Oversight and Accountability, Subcomm. on Economic Growth, Energy Policy, and Regulatory Affairs, 118th Cong. (Jan. 18, 2024). 91 [FOOTNOTE IN ORIGINAL] Katherine Blunt and Jennifer Hiller, Electric Bills Soar Across the Country as Winter Looms, WALL ST. J. (Sept. 18, 2022), available at https://www.wsj.com/articles/electric-bills-soar-acrossthe-country-as-winter-looms-11663493404. 92 [FOOTNOTE IN ORIGINAL] Brian Reidl, The pain isn't going' away: Inflation cost households an extra $10K, N.Y. POST (Dec. 22, 2022) available at https://nypost.com/2022/12/22/the-pain-isnt-goin-away-inflation-costhouseholds-an-extra-10k/. 93 Fueling Unaffordability: How the Biden Administration's Policies Catalyzed Global Energy Scarcity and Compounded Inflation: Hearing Before the H. Comm. on Oversight and Accountability, Subcomm. on Economic Growth, Energy Policy, and Regulatory Affairs, 118th Cong. (Mar. 29, 2023) (statement of Mandy Gunasekara, Director, Center for Energy and Conservation, Independent Women's Forum). Page 19 of 33 "green-at-all-costs" policies aimed at appeasing climate activists at the peril of consumers. Projected demand growth and shifting dynamics in residential, commercial, and industrial sectors are creating massive logistical hurdles for energy regulators while the Biden Administration continues to double down on new regulatory challenges. Indeed, the North American Electric Reliability Corporation noted "Energy Policy" as the number one threat facing our electrical grid in their 2023 report.94

Obama's excursion into green energy venture capital entrepreneurship resulted in many embarrassing collapses.

> Taxpayers had been forced to fund a lengthy string of humiliating failures like Solyndra ($535 million lost), Abound Solar ($400 Million squandered), Range Fuels ($156 million), Evergreen Solar ($25 million), A123 Systems ($279 million), Spectra Watt ($0.5 million), Energy Conversion Devices ($13.3 million), Raser Technologies ($33 million), Mountain Plaza, Inc. ($2 million), Olsen's Crop Service and Olsen's Mills Acquisition Company ($10 million), Thompson River Power LLC ($6.5 million), Stirling Energy Systems ($7 million), Azure Dynamics ($5.4 million), Nordic Windpower ($16 million), Sitcom Technology Corp. ($3 million), Konarka Technologies Inc. ($20 million), Sun Edison ($846 million), Abengoa ($2.9 billion), Beacon Power ($43 million), Ener1 ($118.5 million), Ammonic Solar ($5.9 million), Fisker Automotive ($529 million), Crescent Dunes ($737 million), Willard and Kelsey Solar Group ($0.7 million), etc.

These are just the taxpayer loan recipients who went bankrupt. A lengthier list of government-backed loan and guarantee recipients either failed to produce any tangible value before abandoning their efforts or severely underperformed on their contractual promises.

Green Job Counting

Shortly after the American Recovery and Reinvestment Act of 2009 (ARRA) was signed into law, the U.S. Department of Labor initiated a Green Goods and Services (GGS) survey to demonstrate that the promised bounty of green jobs was being harvested. The heavily left-leaning Brookings Institution initiated a parallel effort, compiling a lengthy report in 2011 entitled Sizing the Green Economy, designed to count the number of green jobs in the economy. It's abundantly clear that the vast majority of green jobs identified by the Brookings authors had long-predated the ARRA "stimulus" plan, Obama's election, or even Obama's birth. Those jobs involved mass transit workers (350,547 jobs), agriculture conservation (314,983 jobs), waste management and treatment (386,116 jobs), etc. However, a comparatively small number still involved green energy generation, electric vehicles, or energy efficiency improvement. As Brookings Institute analysts Mark Muro, Jonathan Rothwell, and Devashree Saha observed in their 2011 report, "the vast majority of clean economy jobs produce goods or services that protect the environment or reduce pollution in ways that have little to do with energy or energy efficiency."

The first GGS report from the Bureau of Labor Statistics (BLS) established a 2010 baseline of 3,129,112 green jobs in all occupational classifications. 2011 BLS published three quarterly GGS employment reports (Q1 2011, Q2 2011, Q3 2011). Despite over $90 billion in government grants and loan guarantees being disbursed, the third GGS report showed only 9,245 green jobs were up to that point. That finding led to a firestorm of criticism in a GOP-controlled Congress. Even lapdog media outlet accounts had trouble disguising the epic failure.

The hilarious aspect of the GGS kabuki theater was the rules BLS employed to boost the count of green jobs. While the green job definition rules are expansive, they can be summarized under two categories:

1. *Jobs in businesses that produce goods or provide services that benefit the environment or conserve natural resources.*

2. *Jobs in which workers' duties involve making their establishment's production processes more environmentally friendly or using fewer natural resources.*

Despite these reasonably straightforward definitions, BLS found 860,289 government workers in its 2010 baseline count of green jobs, even though those workers produce nothing but red tape. BLS published a list of more than 1,200 occupational classifications adjudged to be included or excluded in their green jobs classifications. Consider some of the Byzantine nuances involved in delineating green jobs as opposed to jobs that don't qualify under this critically important workforce accounting agenda:

* A farmer who grows squash using organic farming that contaminates groundwater is a green job worker but not a farmer who uses inorganic fertilizer even though the yield generated by an inorganic farmer maybe 50% higher than his organic farming counterpart, who thus requires far less cultivable, deforested land area for the same quantity of produce, and whose output is 6 times less likely to contain deadly e coli contamination.

* "Food producers who distribute locally and businesses that purchase locally produced food," according to BLS definitions, are classified as green jobs even though locally-grown food is often far more resource-intensive than alternatives produced in more distant locations that benefit from the economies of specialization, scale, scope, and trade.

* A farmer who grows corn for human or animal consumption is not a green job holder. Still, one who sells his output to an ethanol refinery to be incinerated to produce an energy-inefficient, low-power-density fuel is a green job worker.

* A logger who clear-cuts an old-growth forest for timber intended for CO_2-emitting biomass electricity generation is a green job worker but not a logger in a fast-growing, non-emitting, carbon-sequestering, high carbon-stock conifer farm whose output is used to build a house.

* Workers at fabric coating mills are not green job holders, but their colleagues at carpet, rug, and household textile mills may be.

* A technician who repairs a 1968 model refrigerator is classified as a green job worker, while a person who installs a new model might not qualify even though the new model might consume 70% less energy.

* Coal generation jobs are obviously not green jobs, but neither are those in natural gas generation, including workers involved in retrofitting coal generation stations with natural gas turbines that emit half as much CO_2 as coal per unit of energy produced.

* A worker performing automotive repair holds a green job, but not if he repairs the body, glass, or interior.

* A contractor who repairs your stereo has a green job, but not one who repairs your lawn mower.

* Workers at secular grant-making foundations that fund environmental research and education have green jobs, but religious organizations that do exactly the same thing do not.

* Business associations and professional organizations qualify as green jobs, but labor unions do not.

* A docent in an environmental or science museum has a green job but not one in a history or art museum.

* An antique furniture dealer has a green job, but a conventional furniture dealer does not.

* Transportation program administrators who plan mass transit occupy green jobs, but their colleagues in adjacent cubicles who plan roads and bridges do not.

* A K Street lawyer who lobbies for taxpayer-funded, crony-connected green energy megaflops like Solyndra is a green job worker

while a soup-kitchen worker who helps feed the destitute or a nurse who comforts cancer-stricken patients is not.

* A building inspector who audits for environmental compliance holds a green job, but one who performs fire safety certification does not.

. . . and so on.

Don't try to make any sense of this. It is what it is. If you conclude from the foregoing that anyone who degrades economic performance is a green job worker, you're not terribly mistaken. As Diana Furchtgott-Roth of the Manhattan Institute notes, "The entire exercise is an attempt to justify government initiatives, while in practice doing nothing to make America more efficient." It was merely an effort to show that despite all the colossal failures, spending $150 billion of borrowed money was still a great idea to achieve the lowest sustained economic growth rate since the Great Depression. Now, the Biden administration is proposing to spend $2 trillion of borrowed money to perform precisely the identical illogical exercise, promising to obtain the opposite result.

After several quarters of GGS reporting by BLS that had produced nothing but humiliating results, Obama laughingly ordered a total abandonment of the green jobs counting exercise. He preposterously cited budget sequestration constraints. It was imposed by the Balanced Budget and Emergency Deficit Control Act as a reason.

Nevertheless, the Ahab-like quest goes on elsewhere. Brookings updated its 2011 green jobs sizing exercise in April 2019 because, as they claim, "These counts are vital, confirming the extent of the country's energy evolution." You can't just examine energy output statistics. You need to examine how many antique dealers, organic squash farmers, automotive transmission repair mechanics, window installers, local food growers, K Street green megaflop lobbyists, and other vital occupational classification job holders we have. Don't try to make any sense of that, either.

Brookings helpfully tells us, "The transition to the clean energy economy will primarily involve 320 unique occupations spread across three major industrial sectors: clean energy production, energy efficiency, and environmental management." So, we have a definite number of places to look for green jobs, a much smaller count than the BLS list of more than 1,200 occupational classifications. That's a relief.

Brookings goes on to tell us triumphantly that:

"Workers in clean energy earn higher and more equitable wages when compared to all workers nationally. Mean hourly wages exceed national averages by 8 to 19 percent. Clean energy economy wages are also more equitable; workers at lower ends of the income spectrum can earn $5 to $10 more per hour than other jobs."

Needless to say, that's also good news. They neglected to mention how those wages compare against wages in the industrial jobs they're busy destroying. Let's take a look.

BLS reports that a solar photovoltaic panel installer (Occupational Code 47-2231) can earn a median annual wage of $44,890 in 2019, or $21.58 per hour. That's slightly lower than the median wages reported in the same year from the *Clean Jobs, Better Jobs* report table shown in Part 2. A wind turbine service technician (Occupational Code 49-9081) earns a median annual wage of $52,910, or $25.44 per hour, precisely in agreement with the number reported in the same table.

In contrast, construction and extraction workers in coal mining (Occupational Code 47-0000) can expect to earn a median annual wage of $59,380, or $28.55 per hour. In the oil and natural gas pipeline industry, installation, maintenance, and repair occupations (Occupational Code 49-0000) can expect median annual wages of $64,220, or $30.88 per hour. Wages in oil and gas extraction average $101,925 per year or $47.46 per hour.

Brookings assured us that "Workers in clean energy earn higher and more equitable wages when compared to all workers nationally." We know that green energy wages don't compare well to wages in conventional energy industries. But is their statement even true with respect to "all workers nationally"? Solar photovoltaic installers earn $21.58 per hour, and wind turbine technicians earn $25.44 per hour. Meanwhile, "all workers nationally" averaged $28.36 per hour in December 2019, according to the BLS Payroll Survey (or Current Employment Statistics wage data), far higher than green energy workers. Thus, nothing Biden or his media sycophants say about good-paying green energy jobs is accurate.

New Poll: Americans Aren't Willing to Pay for the 'Green New Deal'—And It's Not Even Close

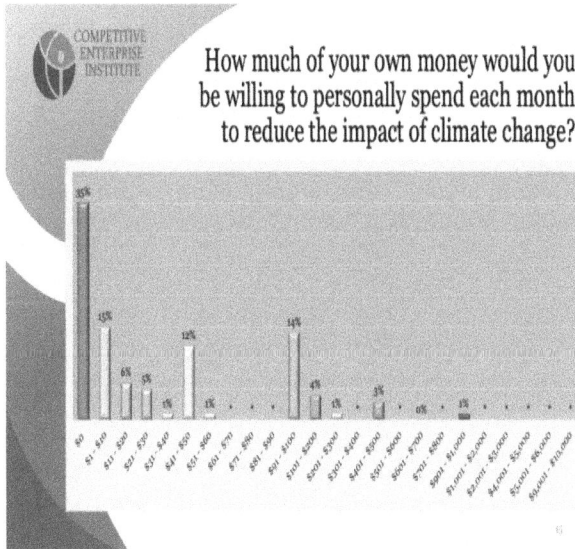

Hopefully, this polling will eliminate ultra-expensive, big-government approaches to addressing climate change.

Americans largely agree that climate change and pollution are real problems. But a new poll reveals that they aren't interested in spending massive amounts on progressive, big-government solutions like the so-called "Green New Deal."

After all, the Green New Deal would cost taxpayers up to $93 trillion, a truly astounding sum of nearly $600,000 per US household. Yet most Americans aren't even willing to sacrifice $50 a month to mitigate climate change. At least, that's the finding of newly released polling from the fiscally conservative Competitive Enterprise Institute (CEI).

CEI surveyed a representative sample of 1,200 registered voters on environmental issues, and their findings have a margin of error of 2.83 percent.

A substantial majority of respondents said they were somewhat or genuinely concerned about the issue of climate change. However, one of the most interesting follow-up questions was this: *"How much of your own money would you be willing to personally spend each month to reduce the impact of climate change?"*

The vast majority of voters were only willing to make very minimal financial sacrifices.

About 35 percent said they wouldn't be willing to spend anything, with another 15 percent saying they'd only sacrifice $1-$10. Another 6 percent were willing to give up $11-$20, while 5 percent said they'd sacrifice $21-$30. In all, 75 percent of respondents were unwilling to pay more than $50 a month.

One need not extrapolate very far from this data to conclude that essentially zero American households are willing to pay $600,000 a year for a "Green New Deal"-style big-government climate change agenda.

"This poll shows once again that Americans are unwilling to pay for the left's anti-energy policies," concluded Myron Ebell, the director of CEI's Center for Energy and Environment. Indeed, it does.

Hopefully, this polling will mean that ultra-expensive, big-government approaches to addressing climate change are taken off the table. From deregulating artificial meat to cutting the red tape blocking emission-free, extremely safe forms of nuclear power, many ways exist to address this issue without digging into Americans' wallets.

Author's Notes:

At this point, I conclude using the many excellent articles written and presented as opinions and findings concerning the New Green Deal (NGD). I will now address the "hot button" areas of the NGD that affect our lives here in the United States like farming, transportation, airlines, infrastructure, electronic cars, trucking, solar, wind power, jobs, housing, carbon dioxide into power, Tokamak Nuclear fusion, Hydrogen power, Hybrid cars, Cleaning oil and gas, turn CO2 into power, California ban on gas equipment, cost/benefit of GND energy policies and a lot more the NGD addresses as issues that need concerned attention.

Our first key article, written by the authors from the Heritage Foundation on July 24, 2019, is one of the first and most comprehensive articles written on the NGD. It draws on 25 supportive contributors from individuals and companies getting to the heart of NGD climate issues. It details key data examples of especially important demographics a person could want to know about the future results we can expect to experience moving through the years.

Green New Deal's Energy Policies

July 24, 2019

Authors: Kevin Dayaratna and Nicolas Loris

Summary

The proposed Green New Deal would be incredibly costly for American families and businesses—for no meaningful climate benefit. Moreover, the plan would introduce a completely new level of cronyism and corporate welfare that would harm consumers multiple times over. The policies proposed in the Green New Deal would disrupt energy markets and skew investment decisions toward politically connected projects. Instead of implementing economically destructive policies of more taxes, regulations, and subsidies, federal and state policymakers should remove government-imposed barriers to energy innovation. Allowing all forms of energy to compete equally in a free market will enable the U.S. to make tremendous strides in terms of a healthy economy as well as a healthy environment.

Key Takeaways

The Green New Deal's government-managed energy plan risks expansive, disastrous damage to the economy, particularly to working Americans.

Under the most modest estimates, just one part of this new deal costs an average family $165,000 and wipes out 5.2 million jobs with negligible climate benefit.

Removing government-imposed barriers to energy innovation would foster a stronger economy and, in turn, a cleaner environment.

On February 7, 2019, Representative Alexandria Ocasio-Cortez (D–NY) and Senator Ed Markey (D–MA) released their plan for a Green New Deal in a non-binding resolution. Two of the main goals of the Green New Deal are to achieve global reductions in greenhouse gas emissions of 40 percent to 60 percent (from 2010 levels) by 2030 and net-zero emissions worldwide by 2050. The Green New Deal's emission-reduction targets are meant to keep global temperatures 1.5 degrees Celsius above pre-industrial levels.

In what the resolution calls a "10-year national mobilization," the policy proposes monumental changes to America's electricity, transportation, manufacturing, and agricultural sectors. The resolution calls for sweeping changes to America's economy to reduce emissions but lacks specific details on how to do so. The Green New Deal also calls for universal health care, guaranteed jobs with a family-sustaining wage, "healthy food security," and efficient spending on all homes and buildings. The analysis in this Backgrounder focuses on the Green New Deal's energy-related policies, which are intended to reduce greenhouse gas emissions.

To provide a broad estimate of the costs, Heritage Foundation analysts modeled the economic impact of an entire series of economy-wide carbon taxes, each increasing the tax gradually over time. We also included regulations and mandates to achieve the

Green New Deal's increased renewable energy generation goal. Our cost estimates constitute a significant underestimate of the true costs of the Green New Deal, as the carbon tax and regulations do not ultimately achieve the policy objectives outlined in the non-binding resolution.

Furthermore, the analysis does not account for the direct taxpayer costs, as advocates want to pay for the Green New Deal through a massive stimulus-style package. Layers of additional regulations and mandates, such as the proposal's objective to maximize energy efficiency for every new and existing building in the U.S., would drive costs even higher. Still, this analysis demonstrates how economically damaging the energy components of the Green New Deal would be for American families and businesses—all for no meaningful impact on the climate.

What Is the Green New Deal?

The Green New Deal is more than just an energy and climate policy; it is a plan to restructure the American economy fundamentally. As stated in the non-binding resolution, "climate change, pollution, and environmental destruction have exacerbated systemic racial, regional, social, environmental, and economic injustices."[1]

To correct those alleged injustices, the plan aims to change how people consume energy, develop crops, construct homes, and produce and transport goods. In other words, the government would use taxes and regulations to control actions and choices made by everyday Americans. Some of the plan's top-line energy goals are to:

- Derive 100 percent of America's electricity from "clean, renewable, and zero-emission" energy sources;[2]
- Eliminate greenhouse gas emissions from manufacturing, agricultural, and other industrial sectors to the extent it is technologically feasible.
- Spend massively on clean-energy manufacturing and renewable-energy manufacturing.
- Eliminate greenhouse gas emissions from transportation and other infrastructure as much as technologically feasible by (among other means) increased government spending on clean infrastructure and high-speed rail;[3]

And Maximize efficiency for every single new and existing residential and industrial building.

What Would a Green New Deal Cost Americans?

Credibly estimating the cost of the Green New Deal's energy policies for American taxpayers, households, and businesses is an exceedingly complex task. The resolution does not specify requiring the grid to transition to 100 percent renewables and instead stipulates "100 percent clean, renewable, and zero-emission" energy sources. How companies would make large-scale investments to meet the mandate and how intermittent power sources would receive backup power is purely speculation and guesswork. Even projecting the cost of switching to 100 percent renewable power for electricity relies on a set of essentially unknowable and untestable assumptions. The costs of stranded assets and lost shareholder value and the cost

---

2 Sixty-three percent of electricity came from CO2–emitting conventional fuels in 2017.

3 Petroleum accounted for 92 percent of America's transportation fuel in 2017. U.S. Energy Information Administration, "Use of Energy in the United States Explained: Energy Use for Transportation," May 23, 2018, https://www.eia.gov/energyexplained/?page=us_energy_transportation (accessed February 1, 2019).

to taxpayers could easily surpass $5 trillion.[4]

Without specific legislative details, assessing the public and private costs is extremely difficult.

To estimate the economic impact of a Green New Deal, we used the Heritage Energy Model (HEM), a clone of the U.S. Energy Information Administration's National Energy Model. As mentioned on Representative Cortez's website, the carbon tax constitutes only one of many policy measures that Green New Deal advocates hope to implement.[5]

As a result, we implemented an economy-wide carbon tax (phased in over two years and increasing by 2.5 percent each year thereafter), a series of regulations on the manufacturing industry encouraging the use of fewer carbon-emitting sources of energy, and mandates for more renewable energy, which currently provides 17 percent of America's electricity needs.[6]

Further details of our modeling are described in the appendix.

The policy's stated goal is to reduce carbon dioxide (CO_2) emissions to zero by the middle of the century, so the first step in our analysis was to ascertain HEM's capabilities for doing so. In particular, we ran a series of simulations with the mandates and regulations described above, gradually increasing the level of the

---

4 Philip Rossetti, "What it Costs to Go 100 Percent Renewable," American Action Forum, January 25, 2019, https://www.americanactionforum.org/research/what-it-costs-go-100-percent-renewable/ (accessed July 10, 2019).

5 Alexandria Ocasio-Cortez, "Green New Deal FAQ," February 5, 2019, https://ocasio-cortez.house.gov/media/blog-posts/green-new-deal-faq (accessed February 7, 2019). As of February 8, 2019, the referenced webpage was no longer available; an archived copy can be found at https://web.archive.org/web/20190207191119/https://ocasio-cortez.house.gov/media/blog-posts/green-new-deal-faq (accessed March 21, 2019).

6 U.S. Energy Information Administration, "Frequently Asked Questions: What Is U.S. Electricity Generation by Energy Source?" March 1, 2019, https://www.eia.gov/tools/faqs/faq.php?id=427&t=3 (accessed July 3, 2019).

carbon tax. Chart 1 illustrates the levels of CO_2 abatement estimated by the model in the middle of the century.

HEM predicts that reducing higher and higher amounts of carbon will not be as simple as instituting higher taxes. Specifically, as the taxes were incrementally increased, the marginal reduction in emissions shrank. In our simulations, a $35 carbon tax results in a 44 percent reduction in CO_2 emissions by 2050, a $100 carbon tax results in a 53 percent reduction, a $200 tax results in a 56 percent reduction, and a $300 tax results in a 58 percent reduction from 2010 levels. Carbon taxes above $300 (resulting in slightly above 50 percent CO_2 reductions by 2050) cause the model to crash, and thus, a 58 percent CO_2 reduction from 2010 levels is the largest level we are able to model.

As a result of the $300 carbon tax, coupled with the regulations and mandates described in the appendix, our simulations find that by 2040, the country will incur the following:

- An overall average shortfall of over 1.1 million jobs.
- A peak employment shortfall of over 5.2 million jobs.
- A total income loss of more than $165,000 for a family of four.
- An aggregate gross domestic product loss of over $15 trillion, and
- Increases in household electricity expenditures averaging 30 percent.

CHART 2

How the Green New Deal Would Affect Employment

The Green New Deal would cause an average annual shortfall of 1.2 million jobs through 2040, with a peak of more than 5.3 million jobs lost in 2023.

CHANGE IN TOTAL EMPLOYMENT, IN MILLIONS OF JOBS

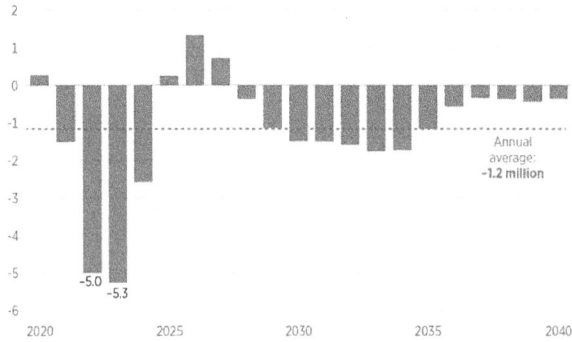

Annual average: –1.2 million

–5.0
–5.3

2020   2025   2030   2035   2040

NOTE: Figures shown are differentials between current projections and projections based on the Green New Deal being enacted in 2020.
SOURCE: Authors' calculations based on Heritage Energy Model simulations. For more information, see the methodology in the appendix.

BG3427 ☎ heritage.org

CHART 3

Family Incomes Would Take Major Hit Under Green New Deal

Under the Green New Deal, the typical family of four would lose an average of nearly $8,000 in income every year, or a total of more than $165,000 through 2040.

CHANGE IN ANNUAL INCOME FOR A FAMILY OF FOUR

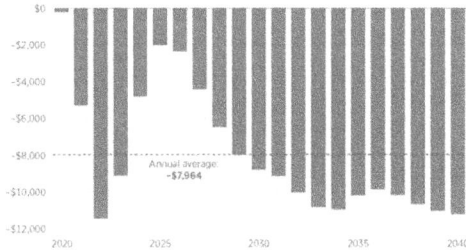

$0

–$2,000

–$4,000

–$6,000

–$8,000

Annual average: –$7,964

–$10,000

–$12,000

2020   2025   2030   2035   2040

NOTE: Figures shown are differentials between current projections and projections based on the Green New Deal being enacted in 2020.
SOURCE: Authors' calculations based on Heritage Energy Model simulations. For more information, see the methodology in the appendix.

BG3427 ☎ heritage.org

167

CHART 5

How the Green New Deal Would Affect Employment in Various Sectors

CHANGES IN EMPLOYMENT BY SECTOR, IN THOUSANDS OF JOBS

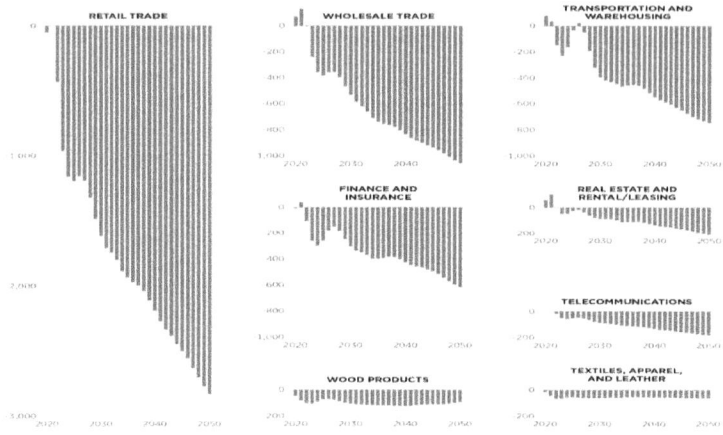

RETAIL TRADE

WHOLESALE TRADE

TRANSPORTATION AND WAREHOUSING

FINANCE AND INSURANCE

REAL ESTATE AND RENTAL/LEASING

TELECOMMUNICATIONS

WOOD PRODUCTS

TEXTILES, APPAREL, AND LEATHER

NOTE: Figures shown are differentials between current projections and projections based on the Green New Deal being enacted in 2020.
SOURCE: Authors' calculations based on Heritage Energy Model simulations. For more information, see the methodology in the appendix.

BG3427 ☒ heritage.org

CHART 4

Green New Deal Would Cause Household Electricity Expenditures to Skyrocket

Under the Green New Deal, household electricity expenditures would rapidly increase by well over 30 percent, and those increases would remain for the foreseeable future.

CHANGE IN HOUSEHOLD ELECTRICITY EXPENDITURES

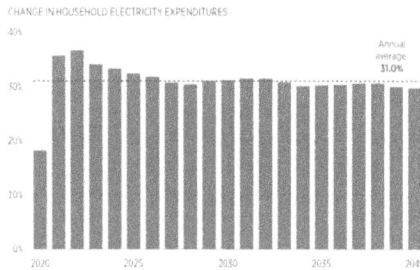

Annual average 31.0%

NOTE: Figures shown are differentials between current projections and projections based on the Green New Deal being enacted in 2020.
SOURCE: Authors' calculations based on Heritage Energy Model simulations. For more information, see the methodology in the appendix.

BG3427 ☒ heritage.org

Chart 5 depicts a sector-by-sector analysis of the impact.

Unquestionably, as the policy only results in 58 percent CO_2 emissions reductions, these estimates significantly underestimate the costs of the Green New Deal. If policymakers spent, taxed, and regulated energy to truly achieve greenhouse-gas-free emission levels, the costs would almost surely be several orders of magnitude higher.

And, more fundamentally, the policies proposed in the Green New Deal are highly regressive. Higher energy costs affect low-income households disproportionately, as they spend a higher percentage of their budget on energy.

What Impact Would a Green New Deal Have on Climate Warming?

No matter where one stands on the urgency to combat climate change, the Green New Deal's policies would be ineffective in abating temperature increases and reducing sea-level rise. In fact, even if the U.S. were to cut its CO2 emissions 100 percent, it would have a negligible impact on global warming. Using the Model for the Assessment of Greenhouse Gas Induced Climate Change, we find that using a climate sensitivity (the warming effect of a doubling of CO2 emissions) larger than that assumed by the Obama Administration's Interagency Working Group, the world would only be less than 0.2-degree Celsius cooler by the year 2100, and sea-level rise would be slowed by less than 2 centimeters.

Chart 6 provides the results from a series of simulations of various climate sensitivities, which demonstrate the negligible climate impact of these policies.

CHART 6

Eliminating All U.S. CO₂ Emissions Would Barely Affect Global Surface Temperatures

Based on various climate model sensitivities.

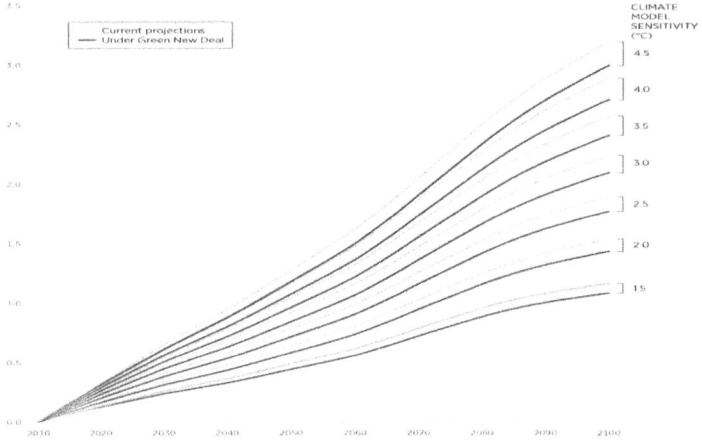

INCREASE IN GLOBAL TEMPERATURES, WITH RESPECT TO 2010 LEVELS, IN DEGREES CELSIUS

Legend:
- Current projections
- Under Green New Deal

CLIMATE MODEL SENSITIVITY (°C): 4.5, 4.0, 3.5, 3.0, 2.5, 2.0, 1.5

SOURCE: Authors' calculations based on Model for the Assessment of Greenhouse-Induced Climate Change (Version 6.0) simulations.

BG3427 ☒ heritage.org

170

Recommendations to Drive Energy and Environmental Innovation

The Green New Deal would amount to more centralization of power in Washington, where the government would determine what type of energy Americans produce and consume. Congress should prevent unelected regulators from misleading the public on the "climate benefits" of greenhouse-gas regulations. Furthermore, policymakers should put forth policy improvements that will drive innovation among all forms of energy. Breaking down barriers to competition, freeing up innovative pathways for innovative technologies, and freely trading energy technologies will meet America's and the world's energy needs while helping the environment. Specifically, Congress should:

- **Require any greenhouse gas regulation to include a separate global-temperature impact and a sea-level-rise impact.** If the purpose of climate-change regulation is to slow warming, then regulators should measure the benefits through the regulation's project impact on warming rather than aggregate emissions reduced, which mislead the public about the benefits of the policy.[7]

The Model for the Assessment of Greenhouse Gas Induced Climate Change provides more useful information for regulators, Congress, and the public when assessing the climate benefits of greenhouse-gas regulation.

- **End the use of the social cost of carbon (SCC) in cost-benefit analyses.** Congress should prohibit any agency from

---

7 See, for example, Kevin D. Dayaratna, Nicolas D. Loris, and David W. Kreutzer, "Consequences of Paris Protocol: Devastating Economic Costs, Essentially Zero Environmental Benefits," Heritage Foundation Backgrounder No. 3080, April 13, 2016, https://www.heritage.org/environment/report/consequences-paris-protocol-devastating-economic-costs-essentially-zero.

using regulatory analysis metrics with the SCC or the "social cost" of other greenhouse-gas emissions in any cost-benefit analysis or environmental review. As has been extensively documented in research by Heritage Foundation analysts, the statistical models on which the federal government relies to estimate the so-called social cost of greenhouse gases are highly prone to user manipulation and are thus not credible tools for policymaking.[8]

If federal courts force regulators into estimating the costs of climate change. In that case, they should not use SCC but the Model for the Assessment of Greenhouse Gas-Induced Climate Change to calculate the global temperature change of regulations or new infrastructure, as has been done in this *Backgrounder*.

- Restate and clarify in law that the Clean Air Act was never

---

8 See, for example, Kevin D. Dayaratna and David W. Kreutzer, "Loaded DICE: An EPA Model Not Ready for the Big Game," Heritage Foundation Backgrounder No. 2860, November 21, 2013, https://www.heritage.org/environment/report/loaded-dice-epa-model-not-ready-the-big-game; Kevin D. Dayaratna and David W. Kreutzer, "Unfounded FUND: Yet Another EPA Model Not Ready for the Big Game," Heritage Foundation Backgrounder No. 2897, April 29, 2014, https://www.heritage.org/environment/report/unfounded-fund-yet-another-epa-model-not-ready-the-big-game; Kevin D. Dayaratna and Nicolas D. Loris, "Rolling the DICE on Environmental Regulations: A Close Look at the Social Cost of Methane and Nitrous Oxide," Heritage Foundation Backgrounder No. 3184, January 19, 2017, https://www.heritage.org/energy-economics/report/rolling-the-dice-environmental-regulations-close-look-the-social-cost; Kevin D. Dayaratna, "At What Cost? Examining the Social Cost of Carbon," testimony before the Subcommittee on Environment and Oversight, Committee on Science and Technology, U.S. House of Representatives, February 27, 2017, https://docs.house.gov/meetings/SY/SY18/20170228/105632/HHRG-115-SY18-Wstate-DayaratnaK-20170228.pdf (accessed July 8, 2019); Kevin D. Dayaratna, "The Flimsy Statistical Models Obama Administration Used to Justify Environmental Agenda," The Daily Signal, March 29, 2017, https://www.dailysignal.com/2017/03/29/the-flimsy-statistical-models-obama-administration-used-to-justify-environmental-agenda/; and Kevin D. Dayaratna, Ross McKitrick, and David W. Kreutzer, "Empirically-Constrained Climate Sensitivity and the Social Cost of Carbon," Climate Change Economics, Vol. 8, No. 2 (April 2017), https://www.worldscientific.com/doi/abs/10.1142/S2010007817500063 (accessed July 10, 2019).

intended to regulate greenhouse gases as air pollutants. Since conventional carbon-based fuels provide approximately 80 percent of America's energy needs, climate-change regulations will drive electricity bills and gas prices higher. Cumulatively, they will cost hundreds of thousands of jobs and tens of thousands of dollars in lost household income and produce no discernable climate benefit.

- **Fix the regulatory and policy problems facing commercial nuclear power.** Facing a complex and burdensome regulatory system, commercial nuclear power in the U.S. has unnecessarily high construction costs. The regulatory system that licenses and permits nuclear reactors fails to keep up with technological innovations and overregulates existing nuclear technologies. Congress should instill regulatory discipline in the Nuclear Regulatory Commission (NRC), encourage the Environmental Protection Agency to right-size radiation-exposure standards, review foreign ownership caps, reform the NRC's cost-recovery structure, and introduce market principles into spent-fuel management.

- **Fix the regulatory and policy problems facing renewable energy.** Like most other energy projects, renewable power projects face excessive and duplicative regulations that increase costs and cause unnecessary delays. Siting and permitting issues can be particularly problematic for wind and solar energy because the most advantageous locations for generation are in more remote areas. Congress should reform outdated environmental statutes, such as the National Environmental Policy Act and the Endangered Species Act, to create a more efficient permitting process for all energy projects, including renewables.[9]

- **Expand energy innovation internationally.** Congress and the Trump Administration should work with other countries to open up their energy markets. These reforms should include pursuing a zero-tariff policy, engaging in technology transfer to unlock natural resources in other countries, and engaging in commercial nuclear trade that would incentivize both cooperation and competition, bringing new nuclear technologies to the market.

Green New Deal: More about Government Control than Climate Control

A Green New Deal would be incredibly costly for American families and businesses—all for no meaningful climate benefit. Moreover, the plan would introduce an entirely new level of cronyism and corporate welfare that would harm consumers multiple times over. The policies proposed in the Green New Deal would disrupt energy markets and skew investment decisions toward politically connected projects, as has been the case with politically favored energy projects in the past.[10]

Instead of implementing economically destructive policies of more taxes, regulations, and subsidies, federal and state policymakers should remove government-imposed barriers to energy innovation. Allowing all forms of energy to compete equally in a free market will enable the U.S. to make tremendous strides in terms of a healthy

---

9 Nicolas Loris and Katie Tubb, "Regulatory Reform Is the Key to Unlocking Infrastructure Investment," Heritage Foundation Backgrounder No. 3384, February 5, 2019, https://www.heritage.org/government-regulation/report/regulatory-reform-the-key-unlocking-infrastructure-investment.

10 See, for example, Nicolas D. Loris, "At Solyndra, More Jobs Lost—and Even More Jobs Lost," The Daily Signal, June 14, 2012, https://www.dailysignal.com/2012/06/14/at-solyndra-more-jobs-lost-and-even-more-jobs-lost/.

economy as well as a healthy environment.[11]

Kevin D. Dayaratna, PhD, is a Senior Statistician and Research Programmer at the Center for Data Analysis of the Institute for Economic Freedom at The Heritage Foundation.

Nicolas D. Loris is Deputy Director of, and the Herbert and Joyce Morgan Fellow in, the Thomas A. Roe Institute for Economic Policy Studies of the Institute for Economic Freedom.

Appendix: Methodology

The Heritage Energy Model

The analysis in this *Backgrounder* uses the Heritage Energy Model (HEM), a clone of the National Energy Model System (NEMS) 2018 Full Release.[12]

NEMS is used by the Energy Information Administration (EIA) in the Department of Energy as well as various non-governmental organizations for a variety of purposes, including forecasting the effects of energy policy changes on a plethora of leading economic indicators.

The methodologies, assumptions, conclusions, and opinions in this *Backgrounder* are entirely the work of statisticians and economists in the Center for Data Analysis (CDA) at The Heritage Foundation and have not been endorsed by and do not necessarily reflect the views of the developers of NEMS.

HEM is based on well-established economic theory as well as historical data and contains a variety of modules that interact with each other for long-term forecasting. In particular, HEM focuses on

---

11 Nicolas D. Loris, "Free Markets Supply Affordable Energy and a Clean Environment," Heritage Foundation Backgrounder No. 2966, October 31, 2014, https://www.heritage.org/environment/report/free-markets-supply-affordable-energy-and-clean-environment.

12 U.S. Department of Energy, Energy Information Administration, "The National Energy Modeling System: An Overview," October 2009, http://www.eia.gov/oiaf/aeo/overview/pdf/0581(2009).pdf (accessed April 3, 2013).

interactions among

- The supply, conversion, and demand of energy in its various forms.
- American energy and the overall American economy.
- The American energy market, the world petroleum market, and
- Current production and consumption decisions, as well as expectations about the future.[13]

These modules are the following:

- Macroeconomic Activity Module,[14]
- Transportation Demand Module,
- Residential Demand Module,
- Industrial Demand Module,
- Commercial Demand Module,
- Coal Market Module,
- Electricity Market Module,
- Liquid Fuels Market Module,
- Oil and Gas Supply Module,
- Renewable Fuels Module,
- Natural Gas Market Module and
- International Energy Activity Module.

---

13 Ibid., pp. 3 and 4.

14 HEM's Macroeconomic Activity Module uses the IHS Global Insight Model, which is used by government agencies and Fortune 500 organizations to forecast the effects of economic events and policy changes on notable economic indicators. As with NEMS, the methodologies, assumptions, conclusions, and opinions in this Backgrounder are entirely the work of CDA statisticians and economists, and have not been endorsed by, and do not necessarily reflect the view of, the owners of the IHS Global Insight model.

HEM is identical to the EIA's NEMS with the exception of the Commercial Demand Module. The Commercial Demand Module makes projections regarding commercial floor-space data of pertinent commercial buildings. Other than HEM not having this module, it is identical to the NEMS.

Overarching these modules is an Integrating Module, which consistently cycles, iteratively executing and allowing these various modules to interact with each other.

Unknown related variables, such as a component of a particular module, are grouped together, and a pertinent subsystem of equations and inequalities corresponding to each group is solved via various commonly used numerical analytic techniques, using approximate values for the other unknowns. Once a group's values are computed, the next group is solved similarly, and the process iterates. After all group values for the current cycle are determined, the next cycle begins. At each particular cycle, a variety of pertinent statistics is obtained.[15]

HEM provides several diagnostic measures based on differences between cycles to indicate whether a stable solution has been achieved.

This *Backgrounder* uses HEM to analyze the impact of carbon tax and carbon-related regulations on the economy. Chart 1 of this Backgrounder illustrates that we modeled $35, $54, $75, $100, $200, and $300 carbon taxes (per ton of carbon). The carbon tax begins in 2020, with half of the specified value per ton of CO2, doubles to its full value the following year, and increases annually by 2.5 percent each year thereafter. In our simulations, each consisting of four cycles, we rebated the revenue collected from the tax back to

---

15 Steven A. Gabriel, Andy S. Kydes, and Peter Whitman, "The National Energy Modeling System: A Large-Scale Energy-Economic Equilibrium Model," Operations Research, Vol. 49, No. 1 (January–February 2001), pp. 14–25, http://pubsonline.informs.org/doi/pdf/10.1287/opre.49.1.14.11195 (accessed December 23, 2014).

consumers in a deficit-neutral manner.

We also implemented regulations on the manufacturing industry by more rapidly retiring CO2-intensive technologies as well as discouraging their use. Lastly, we required that renewable forms of energy constitute a much more significant fraction of the energy portfolio than is currently the case, stipulating that at least 20 percent of renewable electric generation in 2020 come from particular renewable forms of energy and have this percentage gradually increase to 64 percent in 2050. The specific forms of renewable energy we mandated in our simulations included biomass, geothermal, wind, solar, and other forms of intermittent energy.

Using data from the Environmental Protection Agency, we found that the United States emitted approximately 40 percent of all CO2 emissions among all Organization for Economic Co-operation and Development (OECD) member nations.[16]

In our simulations, we altered OECD projections accordingly, assuming this fraction to be constant over time. We also assumed climate sensitivities varying between 1.5 degrees Celsius and 4.5 degrees Celsius, encompassing the "likely" sensitivities specified in the IPCC's "Fifth Assessment Report."[17][25.]

The upper bound of this range is significantly higher than assumed by the Obama Administration's Interagency Working Group in its analysis.[18]

---

16 U.S. Environmental Protection Agency, "Inventory of U.S. Greenhouse Gas Emissions and Sinks 1990–2017," April 12, 2019, https://www.epa.gov/ghgemissions/inventory-us-greenhouse-gas-emissions-and-sinks-1990-2017 (accessed July 10, 2019).

17 U.N. Intergovernmental Panel on Climate Change, "Climate Change 2014," Synthesis Report, 2015, p. 62, https://www.ipcc.ch/site/assets/uploads/2018/02/SYR_AR5_FINAL_full.pdf, (accessed July 16, 2019).

18 Interagency Working Group on Social Cost of Carbon, "Technical Update of the Social Cost of Carbon for Regulatory Impact Analysis."

Authors

Kevin Dayaratna Chief Statistician, Data Scientist, Senior Research Fellow

Nicolas Loris Former Deputy Director, Thomas A. Roe Institute

Author's Notes:

I don't know about any of you who have read and analyzed the presented data and narrative, but I cannot agree that the materials Kevin and Nicolas presented and detailed to a high degree of acceptance and pertinent applicability have served us well. I congratulate them and the Heritage Foundation for producing such a well-accepted article.

Innovation Exists and May Be a CO2 Solution!

Let's see what this Italian company is doing and PST! US companies, please look at this for hints, ideas, and actionable intercontinental collaboration. This looks good to me.

"A giant energy dome is daringly turning carbon dioxide into power,"

— Daniel Grizelj

- One of the major engineering challenges facing the green energy revolution is the need for cost-effective methods of storing energy.

- Energy Dome, an Italian startup, is turning to CO_2, the leading culprit of the climate crisis, to try to solve this lasting conundrum by compressing the greenhouse gas into a liquid for storage and eventually using that gas to power turbines when the sun isn't shining, and/or the wind isn't blowing.

- While this has some big advantages over similar techniques, particularly because CO_2 can stay liquid at ambient temperature (under high pressure) and is more energy-dense than air, the technology could eventually become outclassed by the rapidly improving battery technology.

One of the big obstacles between today and our green energy future is energy storage. After all, sometimes the wind doesn't blow, and the sun doesn't shine, so designing ways to store energy effectively is an extra-critical component of combating climate change. That's why engineers have developed new types of batteries, whether powered by lithium, rust, or even gravity, to figure out ways to tackle this relentlessly difficult engineering problem.

Now, the Italian startup *Energy Dome* is ready to debut its counterintuitive approach to energy storage: combating decarbonization with … more carbon.

For years, the company has perfected the chemical technologies needed to create a carbon dioxide-powered battery. Because CO2 liquifies at ambient temperatures under pressure and has higher energy density than air, this approach has some clear advantages over similar techniques such as liquid-air energy storage and compressed-air energy storage.

This 'Cage of Cages' Stores CO2 Faster Than Trees

As the company's name suggests, 1000 tonnes of CO2 gas is first stored in a giant dome. While solar panels and wind turbines are pumping out electrons, Energy Dome pumps the gas into a compressor, where it's heated, compressed into a liquid, and stored in carbon steel tanks. Then, when energy is inevitably needed during those sunless and/or windless moments, an evaporator turns the liquid into pressurized gas, and once reheated, the released CO2 powers turbines and generates electricity.

According to a 2022 interview with MIT Technology Review, Spadacini says a full-scale plant should cost $200 per kilowatt-hour (kWh). In contrast, lithium-ion battery tech currently costs $300 per kWh. *New Scientist* reports that such a plant will store power for 8 to 24 hours.

"What we provide is a technology which is intended to be in the daily storage market. It means to switch energy from day to night, from today to tomorrow," Spadacini told *New Scientist*. "That is your main need with a wind and solar-dominated grid."

So, could the object of our greenhouse gas derision somehow provide the silver bullet necessary to slay this energy storage lycanthrope? The idea certainly has some big backers, as Energy Dome receives funds through Bill Gates' climate investment initiative, Breakthrough Energy Catalyst, and the European Investment Bank.

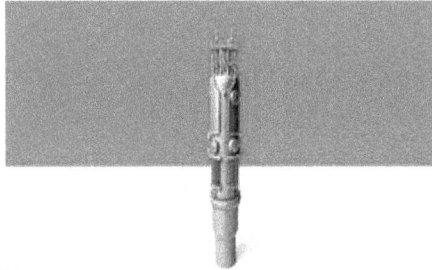

Now, after years of discussing this, Energy Dome is implementing its CO2-powered battery plan with the first operating commercial plant due to open in Sardinia, Italy, sometime later this year. The good news doesn't stop there; Spadacini also says that "a very large, well-known global utility" is interested in the technology.

Of course, downsides exist. For one, a giant, big ol' bag of still-air CO_2 could be a problem if somehow ruptured, and these plants aren't exactly small; *New Scientist* writes that a CO_2 battery plant could assume anywhere from 12 to 25 acres, or roughly a 3 to 4 megawatt solar farm.

Do you know what the other problem is? Lithium-ion batteries, as well as other next-gen battery technologies, are getting better and better. However, they are already mass-produced, and another problem is that. Lithium-ion batteries are relatively modular, meaning they're a bit more plug-and-play than a massive CO_2 battery plant.

However, the need for cost-effective energy storage has never been greater and will only increase as renewable energy supplants the fossil fuels that got us into this mess in the first place. If CO_2 can do good rather than harm, then let's put it to work.

Alliant Energy utility wants to demonstrate the nation's first CO2-based long-duration 'Energy Dome.'

"The unique energy storage technology could approach a round-trip efficiency of up to 75%," says an Alliant official.

Published Aug. 20, 2024

By

Emma Penrod

A rendering of the Columbia Energy Storage Project, a 20-MW/200 MWh energy storage system Alliant Energy and other utilities plan to build near Portage, Wisconsin.

Courtesy of Alliant Energy

Dive Brief:

- Alliant Energy utility Wisconsin Power and Light, Madison Gas and Electric Co., and Wisconsin Public Service Corp. have requested approval from the Public Service Commission of Wisconsin to construct the Columbia Energy Storage Project, an up to 20-MW carbon dioxide-based energy storage system near Portage, Wisconsin.

- The project will evaluate a new long-duration storage concept pioneered by Italian startup Energy Dome. The concept involves using compressed carbon dioxide to drive an otherwise conventional turbine. The company says the technology could achieve a round-trip efficiency near 75% without needing lithium or other critical minerals.

- The demonstration project aims to evaluate how best the new storage technology might fit within the company's grid operations, said Mike Bremel, director of technical solutions and federal funding at Alliant Energy.

WOW! And are the NGD folks not even considering this? Do you ever wonder why?

Opinion

By The Latest Magazine Podcast Archive

Be Careful What Energy You Wish For

Debate: Green New Deal vs. Degrowth, Part 2

Don Fitz

October 15, 2021

This is the second part of an exchange about the climate crisis and a global green New Deal vs. Degrowth as workable solutions between Robert Pollin, a world-renowned progressive economist, and Don Fitz, a social psychologist, author, and Green Party leader. The full exchange originally appeared in ZNet. Part 2 of the debate originally appeared in Green Social Thought. Read the rest of the debate.

Climate change is an existential problem confronting humanity and Robert Pollin offers well-thought-out plans in his first and second articles. There is only space here to respond to these themes: (1)

We can reach zero CO_2 emissions by 2050.

An enormous expansion of alternative energy (AltE) can create far more jobs than would be lost by eliminating fossil fuels (FF) and

Without stating so explicitly, he implies that the negative aspects of AltE would be tiny.

The downsides of AltE are so significant that they must form the core of any serious discussion about our response to the climate crisis.

Unfortunately, these drawbacks are so large that they must be central to any meaningful discussion. Like many proponents of AltE, Pollin frames climate change not as a consequence of capitalist overproduction but as a stand-alone crisis whose resolution may require strengthening capitalism. Yet, there is already more than enough manufactured production to meet the needs of humanity if only manufacturing were shifted from creating what is wasteful and destructive to production for need.

The rush to massively expand AltE ignores what environmentalists have known for decades. Redesigning cities to create walkable communities is vastly more ecological than subjecting children in the Democratic Republic of Congo to the horrors of cobalt mining for electric vehicles. Humanity can reduce its footprint by consuming much less meat and relying on local food rather than transporting the average bite 1,700 miles. Americans do not need housing space vastly exceeding that of anywhere else in the world (except Australia). Buildings can be altered for reuse instead of constant tear-down rebuilding. Manufacturing standards can be set to require the maximum life for goods rather than subjecting people to the message of "Buy, Buy, buy, buy!" as corporations design products to fall apart and go out of style. Humanity does not need more obstacles — corporations do.

Rather than exploring these, Pollin labors to satisfy the corporate urge for infinite growth via "clean, renewable" energy sources of solar, wind, "small-scale hydro," and "low-emissions bioenergy." It is also necessary to note that achieving a "zero emissions economy" by 2050 is a disingenuous goal. The life cycle of AltE requires FFs for construction equipment, mining for essential minerals, transportation, and of disposal of energy harvesting equipment. Wind and solar are intermittent, requiring backup (often gas) for emergencies.

Solar Power

In addition to health problems from extraction, manufacturing, and disposal, there are issues of those who would be affected. Greg

Wagner recently wrote me: "I live in a small rural farming community known as Goldendale, Washington, in Klickitat County on the east side of the state. I formed the group Citizens Educated about Solar Energy (CEASE) when I learned by accident the solar companies' plans. My home and many others are surrounded by tens of thousands of acres of wheat/alfalfa/hay fields and beautiful views of Mt. Adams, Mt. St. Helens, and Mt. Hood. This will all be destroyed if our county commissioners allow Invenergy and Cypress Creek Renewables to build their proposed solar sites on the 10,000 acres we know about…"

On the other side of the US, citizens in Culpepper County, Virginia, blocked a proposal by Maroon Solar. The planning board said that "the project was out of character with the area, too large, and would be injurious to the health and safety of area residents." Opponents also commented that "the utility-scale project would ruin the rural nature of the area, the environment, and historic resources."

Wind Power

Health problems also appear with wind, and victims of its placement are often unhappy. Opponents halted wind turbines in Reno County, Kansas. As in other areas, a major concern was having wind turbines too close to people. As Patrick Richardson reported, "The setback is often no more than double – sometimes just one and a half times – blade height, meaning turbines can be within as little as 500 feet away from a school, as in Labette County."

Showing that the problem is international, Shivani Gupta writes of 1000 villagers in the Kachchh district of India who protested plans to put industrial wind turbines in portions of the Sangnara forest and adjacent common grazing areas where farmers bring their cows and sheep. "The Sangnara forest is home to a huge diversity of endangered flora and fauna. This includes chinkara, wolf, caracal, ratel, hyena, desert cat, Indian fox, spiny-tailed lizard, desert monitor, white-naped tit, vultures, and many more."

186

"Small Scale" Hydro

Such power comes from dams with a capacity of less than 30 MW. While these dams are individually less destructive than mega-dams, a much, much larger number would be needed to provide power for an incessantly growing capitalism.

At 22 MW capacity, the Agua Zarca hydro-project in Honduras was "small-scale." Berta Cáceres won the 2015 Goldman Environmental Prize for organizing the Lenca people, along with other Indigenous groups and the Afro-Honduran Garifunas, to oppose the dam. The following year, she was murdered by those who stood to profit from "clean, renewable energy." In addition to being sacred, "the Gualcarque River is a primary source of water for them to grow their food and harvest medicinal plants. Dams can flood fertile plains and deprive communities of water for livestock and crops."

"Low-Emissions" Bioenergy

Burning organic material is now the rage in Europe. In April 2021, climate scientists James Dyke, Robert Watson, and Wolfgang Knorr, who had formerly supported plans for "net-zero" emissions, did an about-face as they wrote about its latest version, Bioenergy Carbon Capture, and Storage, or BECCS: "Rather than allow ecosystems to recover from human impacts and forests to regrow, BECCS generally refers to dedicated industrial-scale plantations regularly harvested for bioenergy rather than carbon stored away in forest trunks, roots, and soils. Currently, the two most efficient biofuels are sugarcane for bioethanol and palm oil for biodiesel – both grown in the tropics. Endless rows of such fast-growing monoculture trees or other bioenergy crops harvested at frequent intervals devastate biodiversity. It has been estimated that BECCS would demand between 0.4 and 1.2 billion hectares of land. That's 25 percent to 80 percent of all the land currently under cultivation."

BECCS can also include burning wood pellets. Land Institute research scientist Stan Cox observes that "This process is not emissions-free; in fact, wood-fired power plants produce more pollution per kilowatt-hour of electricity than coal-fired plants." Even if BECCS were used to capture those emissions, Cox notes that "The process's many steps – producing and harvesting biomass crops, hauling the biomass to the processing factory, grinding and pelletizing, hauling pellets to the power plant, sucking carbon dioxide out of the smokestack, liquefying the carbon dioxide, hauling the liquid to the injection field, and shooting it into the earth under high pressure – would, in sum, eat up large quantities of energy."

Hey, has anyone thought about renewable energy? Well, let me tell you about it. First off—and just a disclaimer that, but I do not know this for sure—I can believe ancient people going back to ancient Africa and the first people using wood fires to heat the caves and to provide light would be a qualifier. Does it have a place in today's world? You bet on heating, artificial lighting in buildings, and a lot more.

I have two articles I will present to you. One is 2024, which gives us the outlook for renewable energy, and another on underground renewable energy. Then, one on an experimental called the "Holy Grail," of renewable energy as a functioning. I know I sound like a naysayer of the NGD, but don't you think that with all that need for scads of money and massive programs, why can't we pause, breathe, and create using age-old technology to beat most of the problems?

Okay, let's level the playing field. We cannot, and I repeat, we CANNOT conquer those things the Earth has been doing for billions of years. It will do as it wishes, and we have to listen and try to survive. Yes, survive, just like the Earth has. Can you even imagine this? Can you tell me how many times the Earth has reset itself? Dinosaurs are gone, floods are occurring, a whole lot of people are vanishing as if they never existed, have vanished like a star, and diseases are wiping lives off. These are some bizarre occurrences we cannot really explain, and some things I cannot myself fathom why they happened.

We do have elected officials to deal with such or similar circumstances. However, those officials cannot and will not conjure up the actual cause as to why they have an Earth crisis or identify which ones we can control, and which ones are beyond control. If we pay them enough money, chances are they will try.

Ancient Greeks and Romans used wind-powered water pumps to irrigate their crops, and ancient Chinese and Persians used water wheels to grind the grain. The earliest known use of solar energy was grain. In ancient China, mirrors, now called 'burning mirrors,' were used to light fires for religious ceremonies.

Ancient Civilization

The use of solar energy can be traced back to ancient civilizations, where people relied on the sun for warmth, light, and food preservation. One of the earliest known uses of solar energy was in ancient China, where mirrors, now called 'burning mirrors,' were used to light fires for religious ceremonies as early as 4000 B.C.

In ancient Greece, the philosopher Socrates observed that south-facing buildings received more sunlight in winter, leading to the design of passive solar architecture. Ancient Egyptians also utilized solar energy, aligning their buildings to capture the sun's rays for heating and lighting. These early civilizations recognized the sun's power. They sought to use it for their daily needs, laying the foundation for future solar technology development.

Early Solar Technologies

In the 7th century AD, the magnifying glass, used to start fires by concentrating sunlight, was a precursor to modern solar cookers. In the 18th century, Swiss scientist Horace-Bénédict de Saussure built the first solar oven, which reached temperatures above 446 degrees Fahrenheit (230 degrees Celsius).

Around the same time, 19-year-old French physicist Alexandre-Edmond Becquerel discovered the photovoltaic effect. He noticed that when electrons in a material's conduction band became excited, they could move freely, generating an electric current. This laid the foundation for solar photovoltaic (PV) technology, which is the first solar cell.

In the 1860s, French mathematician August Mouchot envisioned solar-powered steam engines. Over the next two decades, Mouchot and his assistant, Abel Pifre, brought this idea to life by constructing the first solar-powered engines used for various applications. This pioneering work laid the foundation for modern parabolic dish collectors.

In 1873, Willoughby Smith made a significant discovery regarding selenium's photoconductive potential. This led to William Grylls Adams and Richard Evans Day discovering in 1876 that selenium could generate electricity when exposed to sunlight.

Charles Fritts is credited with building the first true solar cell in 1883. He coated selenium with a thin layer of gold, forming a semiconductor junction that produced electricity when exposed to light. Fritts' invention was the precursor to modern solar cells, marking a significant advancement in solar technology.

Trivia: Did you know that in the 1700s and 1800s, sunlight was harnessed to power ovens and even ships? This early use of solar energy demonstrates humanity's long history of utilizing the sun's power for practical purposes.

Solar Water Heating and Other Breakthroughs

The 19th century saw significant advancements in solar energy technology, particularly in solar water heating. In 1891, American inventor Clarence Kemp patented the first commercial solar water heater used in homes and businesses to heat water using sunlight. This innovation laid the groundwork for modern solar water heating systems.

In 1908, William J. Bailley of the Carnegie Steel Company invented a solar collector featuring copper coils and an insulated box, a design that resembles present-day solar collectors.

In the mid-1900s, a significant milestone in solar energy was reached in 1954, when Gerald Pearson, David Chapin, and Calvin Fuller developed the first functional silicon solar cell. Their achievement marked a breakthrough, achieving an efficiency of 11% with a cell the size of a small coin. This accomplishment made history as they became the first to convert sunlight into usable energy successfully.

Western Electric began selling commercial licenses for silicon photovoltaic (PV) technologies. These early PV products included dollar bill changers and devices for decoding computer punch cards and tape.

In the mid-1950s, architect Frank Bridgers designed the first commercial office building to use passive design principles and solar water heating. Known as the Bridgers-Paxton Building, this landmark has been continuously operating with its solar system since its construction. It is now listed on the National Historic Register as the world's first solar-heated office building.

Going green: The history of re"new"able energy

Energy and transportation are historically linked. All of the ways that people and commodities go from place to place, require energy, much of which is from non-renewable sources (i.e., coal, oil, and natural gas). These non-renewable sources can be depleted. They can run out. First, there will be shortages. Shortages occur, followed by elevated prices. Eventually, the resource becomes unavailable.

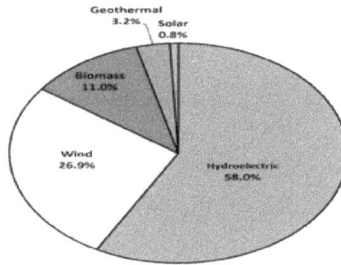

Total percentage of renewable energy used.

in the US. Photo from Wikipedia.

Did you know there is a direct link between energy from non-renewable sources and poor environmental and human health? For example, over five million workdays have been canceled, and 600,000 cases of asthma attacks have been linked directly to pollution from nearby fossil-fuel power plants. Because of this, the United States has recognized the need for a more diverse "energy profile," which means getting energy from different, especially "renewable," energy sources.

But what is "renewable energy?" It is an energy source. Except this one can be used over and over again while naturally regenerating itself. Think of solar power, for example. Solar energy is derived from the sun's radiation. The sun is a powerful source of energy and provides the Earth with as much energy every hour as we collectively use in a year worldwide. Some other examples of renewable energy sources used in the United States include wind, geothermal, biomass, and hydropower.

Renewable energy has been around for forever, right? Correct! The history of renewable energy sources dates back to the beginning of civilization. However, only in recent decades has technology been a primary focus as the United States over energy shortages and the prospect of climate change continues to build.

When did it all start?

Solar energy

Did you know that William Robert Grove was the first person to discover how to harness solar energy in 1839? He was also the inventor of the first hydrogen fuel cell 1. However, it took almost 100 years before his technology became 10 percent efficient, making it more attractive for implementation. A large-scale solar plant was constructed in California in 1981, utilizing 1,818 mirrors to reflect light into a receiver that generated heat to operate a generator.

Hydroelectric Power

Did you know that the earliest people to utilize water for power were the Europeans in 200 BC? They used this water power to control mills that crushed grain. Then, after several centuries, water was used to harness the power of electricity. The first hydroelectric power plant was built in Appleton, Wisconsin 1882.

A hydroelectric power plant captures the energy of falling water by storing it in a reservoir. The water is then released by a dam, which can control the flow of water falling (i.e., control the amount of energy released). The water turns a turbine, which powers a generator and disperses energy through transmission lines.

Hydroelectric power stations are still widely used, with the highest plant concentration on the Columbia River in Canada, Washington, and Oregon. Dam production peaked in the 1960s, and then, following environmental legislation by the federal government, the creation of new hydroelectric power plants decreased due to their negative impact on water ecosystems.

Wind Energy

Did you know windmills were initially used to grind grain beginning as early as 9,000 AD in the Persian Empire (Middle East)? Later, in the 1590s, the Dutch adopted windmills to pump water and reduce flooding. It was not until 1888, in Cleveland, Ohio, that the first wind turbine was created to generate electricity. Today, the largest wind farm in the world is located in California, which includes 6,000 wind turbines over 50,000 acres. Why so many? Construction of wind farms began in 1981 in response to the energy crisis in the 1970s, which prompted the Department of Energy to fund major wind turbine designs, making large-scale wind farms possible.

Biomass

Did you know that the alcohol you drink (i.e., Ethanol powered engines as early as 1826. The first ethanol plant was built in the 1940s. In the United States, most ethanol is actually created through the fermentation of corn. During the Great Depression, from 1929 to 1939, ethanol production policies were enacted to support corn production and stabilize corn prices for farmers.

However, the production and consumption of ethanol really "powered up" in 2007 when the United States passed the Energy Independence and Security Act, which put a minimum Renewable Fuel Standard (RFS) of 15 billion gallons by 2015. The RFS is a federal program that requires that at least 10 percent of transportation fuel originate from renewable energy, such as ethanol.

Geothermal Power

Did you know that the world's first known geothermal (i.e., using the heat of the Earth to create energy) district heating system was created in Idaho in 1892? It was only in 1921 that the inaugural geothermal power plant was constructed in California, marking a significant milestone in renewable energy development. Geothermal power really took off in 1970 following the implementation of the Geothermal Steam Act, which allowed the leasing of federal land (other than land in National Parks) that contained geothermal resources. Today, the Bureau of Land Management manages 818 geothermal leases, in the western United States and Alaska, creating enough power to satisfy electrical needs for 1.5 million homes.

Why is all the interest?

Did you notice anything in common between each renewable energy source (besides hydroelectricity)? These sources increased drastically within the last few decades, beginning in the 1970s. What was the motivation? Most people point to the oil crisis of 1973 and 1979 as the impetus for the United States to think more seriously about renewable energy.

An energy crisis happens when demand outweighs supply (whether due to infrastructure issues, limits in production, government tax hikes, market monopoly, or other reasons). This crisis had a significant impact on the price of energy. As an example, let's take the oil crisis of 1973. The reason for the energy crisis was due to an embargo (ban on trade) against the United States from oil-rich countries due to international conflict. The involvement of the United States in the Yom Kippur War created tension with oil-rich countries like Saudi Arabia, Kuwait, Iraq, and Qatar. As a result, oil prices rose 300 percent in just one year.

The effects of this on the transportation sector were huge. Cars became much more fuel efficient, averaging 17.5 miles/gallon in 1985 compared to 13.5 miles/gallon in 1970. Also, there was a greater focus on harnessing renewable energy, specifically wind and solar energy. Additionally, interest in mass transit increased greatly following the oil crises. Since 1972, the number of riders using public transit has grown by 55 percent, making the number of riders over 10.4 billion today.

Is alternative energy still popular?

The increase in renewable energy clearly coincides with the energy crises of the 1970s. The United States has begun investing heavily in alternative energy sources to diversify its energy portfolio and offset any impacts from future energy crises, which are likely to happen. However, researchers from the Information Technology and Innovation Foundation found that the United States should spend between $8-30 billion per year on renewable energy research to address climate change. In contrast, actual spending is around $3.5 billion per year. Today, the United States consumes almost 20 percent of the world's energy, but it only makes up five percent of its population. Reducing our reliance on energy through efficiency and conservation will help, but we also need to invest in renewable energy sources for our future.

This next example paper is from Deloitte Insights, and it covers the outlook for renewables with investment, competitiveness, demand, supply chain, and workforce challenges. As the author, I cannot say much about this article since it is quite detailed and self-contained.

The solar market brightened in 2023 in a bifurcated renewable landscape, while wind faced sweeping challenges. The latter withstood the worst of project inputs, labor and capital cost pressures, interconnection and permitting delays, and transmission limitations. Meanwhile, supply chain constraints started easing as historic clean energy and climate laws took effect.

In the United States, utility-scale solar capacity additions outpaced additions from other generation sources between January and August 2023—reaching almost 9 gigawatts (GW), up 36% for the same period in 2022—while small-scale solar generation grew by 20%.[1] Only 2.8 GW of wind capacity came online during the same period, down 57% from last year, resulting in renewables accounting for just over half of capacity added versus two-thirds last year. However, renewable energy's share of US electricity generation remained level at 22%. By the end of 2023, the US Energy Information Administration expects utility-scale solar installations to more than double compared to 2022, to a record-breaking 24 GW, and wind capacity to rise by 8 GW.

2024 energy, resources, & industry outlooks

The tandem push of federal investments flowing into clean energy and the pull of decarbonization demand from public and private entities have never been stronger. Moving into 2024, these forces could enable renewables to overcome hurdles caused by the seismic shifts needed to meet the country's climate targets. The uplift and obstacles shaping the year ahead have set the stage for a variable-speed takeoff across renewable technologies, industries, and markets.

Table of contents

Federal investment push

Deployment highs. The Energy Information Administration expects renewable deployment to grow by 17% to 42 GW in 2024 and account for almost a quarter of electricity generation. The estimate falls below the low end of the National Renewable Energy Laboratory's assessment that Inflation Reduction Act (IRA) and Infrastructure Investment and Jobs Act (IIJA) provisions could boost annual wind and solar deployment rates to 44 GW to 93 GW between 2023 and 2030, with the cumulative deployment of new utility-scale solar, wind, and storage reaching up to 850 GW by 2030.

Cost lows. A temporary rise in renewable costs could belie their long-term declining trend and relative competitiveness. High financing, the balance of plant, labor, and land costs outweighed commodity and freight price falls in 2023, pushing up the levelized costs of energy (LCOEs) for wind and utility-scale solar, especially projects with trackers that account for 80% of installed solar capacity. Inflation and interest rates disproportionately impacted offshore wind, which saw a 50% rise in its LCOE from 2021 to 2023.

While this equation may prevent LCOEs from resuming historical downward trends in 2024, the IRA investment tax credits and production tax credits have made utility-scale solar and onshore wind, including projects paired with storage, competitive with marginal costs of existing conventional generation. Projects claiming the maximum available credits could capture the world's lowest solar and wind LCOEs.

Renewables collecting production tax credits will likely increase the prevalence of negative prices in wholesale electricity markets.

Decarbonization Demand Pull

Most states and utilities. Twenty-nine jurisdictions, representing around half of US electricity retail sales, have mandatory renewable portfolio standards (figure 7); 24 jurisdictions, including two new states in 2023, have zero greenhouse gas (GHG) emissions or 100% renewable energy goals spanning 2030 through 2050. Renewable portfolio standards and clean energy standard policies are expected to require 300 terawatt hours (TWh) of additional clean electricity by 2030.[13]

Complementary to state goals are the 56 individual and 28 parent utilities with carbon reduction targets that serve 83% of US customer accounts.[14] Twenty-five utilities have further committed to either an 80% carbon reduction or an 80% share of clean generation by 2030.[15] More states, localities, and public utilities are expected to invest in renewables in 2024, as the IRA's direct pay and transferability mechanisms help enable their market participation.[16]

Major corporations. In the first ten months of 2023, 30 companies joined RE100, a global corporate initiative to procure electricity entirely from renewables, growing the membership to 421.[17] Around a quarter of the members are headquartered in the United States, and a bulk of their upcoming commitments have a 2025 target date. Some are also driving decarbonization throughout their supply chains. Following a record-breaking year, corporate renewable procurement saw the number of transacting customers increase by 31% between the first half of 2022 and that of

2023.[18] Big technology companies accounted for most of the procured capacity[19] a trend likely to grow in 2024 as the companies meet and help others meet 24/7 and carbon-matching targets with the help of generative artificial intelligence.[20] The training and use of generative AI could increase their data center demand for clean electricity five- to sevenfold.[21] A growing number of corporations are also expected to support renewables by participating in the nascent tax-credit transfer market in 2024. Coming full circle, corporations are participating in multinational efforts to push governments to address climate change and accelerate the energy transition. Ahead of COP28, 131 companies with close to US$1 trillion in annual revenue drove a campaign urging governments to phase out fossil fuels by 2035.[22]

The impact of unprecedented investment in renewable infrastructure will likely become more apparent in 2024. Regulatory boosts to renewable energy and transmission buildout could help address grid constraints. And boosts to manufacturing could lay the foundations of a domestic clean energy industry with stronger supply chains supporting solar, wind, storage, and green hydrogen deployment. A skilled workforce should be prepared to build, operate, and maintain all these new generation and manufacturing facilities planned over the next few years. As renewables become a larger part of power generation and the portfolio of technologies grows, perceptions could start catching up with the reality that renewables can enhance grid resilience. Deloitte's 2024 renewables industry outlook discusses how these trends could impact the industry in the coming year:

1. Regulatory Boosts and Brakes: Historic investment could erode obstacles

The IIJA and the IRA have boosted renewables through historic investment in new or expanded programs, grants, and tax credits to accelerate the deployment of established and emerging renewable technologies. Over the past two years, they helped catalyze US$227 billion of announced public and private investments in utility-scale solar, storage, wind, and hydrogen.[23] To date, US$100 billion of these investments have materialized, in addition to US$82 billion in distributed renewables and heat pumps (figure 1). States also offered a record US$24 billion in tax breaks in 2022 to attract projects.[24] The bulk of investment flowed to states with ambitious decarbonization targets and mandates, led by California, as well as states with greater renewable resources and lower permitting and siting costs, led by Texas and Florida.[25] An outsize share of clean energy investment also flowed to energy, disadvantaged, and low-income communities identified in the IRA for additional incentives.[26]

Solar and Storage Soar

The IIJA and the IRA have had some of the biggest impacts on solar and storage. Utility-scale solar captured the largest share of both announced investment of US$92 billion and actual investment of US$52 billion across 38 states. The month after the IRA passed, a record 72 GW of standalone solar was added to the interconnection queue, more than the preceding 11 monthly additions combined.[27] Amid a venture capital (VC) industry slowdown, VC funding for solar and storage increased in the first three quarters of 2023, and the IRA boost blunted higher interest rates as public market and debt financing for solar also grew.[28] Solar recorded 34% growth in actual investment over the past year, storage jumped 51%, and distributed renewables, storage, and fuel cells increased 31%.[29]

IRA tax credits have allowed solar and storage developers to creatively configure projects around sitting and grid constraints through standalone or hybrid deployment. In 2024, tax credit adders are expected to shape solar and storage market offerings.[30] US Treasury's release of guidance on energy and low-income community adders in the last quarter of 2023 could be particularly relevant to community solar developers.[31] The guidance may also drive more third-party-owned solar and storage projects, which can qualify for these adders, unlike customer-owned systems.[32] Finally, the impact of US$7 billion in Greenhouse Gas Reduction Fund grants should be seen through the Solar For All program and the US$3 billion loan from the Department of Energy's (DOE's) Loans

Program Office (LPO), which is the government's largest-ever single commitment to solar.[33] Both programs focus on distributed solar and storage deployment in low-income and disadvantaged communities.

Hydrogen Teed Up for Takeoff

The IIJA and the IRA have teed up the takeoff of a new green hydrogen economy. Clean hydrogen has the largest gap between announced and actual investments—more than US$50 billion and less than US$1 billion, respectively.[34] The gap reflects in part uncertainty over pending Treasury guidance on tax credits that are expected to make green hydrogen competitive. At stake is whether hourly matching, additionality, and deliverability will be required to qualify for the full US$3/kg credit.[35] Favorable final guidance could open the floodgates on actual investments in 2024 and jumpstart the nascent hourly Renewable Energy Credit (REC) market. Treasury alignment with the European Union's approach to gradually phase in the three requirements could also enable the US industry to serve the 10 million metric tons (MMT) clean hydrogen import market that the EU envisions by 2030.[36] Exports could help resolve demand uncertainty reflected in the DOE's reallocation of US$1 billion in hydrogen hub funds to stimulate demand.[37] In 2024, the industry should watch for developments in the seven selected hydrogen hubs as they move into their design and planning phase, as well as the launch of the country's first end-to-end green hydrogen system.[38]

Energy Efficiency Inches Up

Investment following the IRA fell short of meeting ambitious targets for energy efficiency, as reflected in heat pump deployments. While heat pumps attracted close to US$45 billion, investment has only grown 1% over the past year.[39] Final DOE guidance on state administration of direct customer rebate programs could unleash growth in 2024. An initiative of the US Climate Alliance of 25 states to install 20 million heat pumps by 2030 could further bolster deployments,[40] as could utility-funded energy efficiency programs.[41]

Wind Brakes

Wind investment dropped 35% over the past year as projects bore the brunt of headwinds from higher costs and permitting challenges, which respondents of a Deloitte survey identified as the most significant constraints on renewables (figure 2).[42] Record curtailments felled wind in most independent system operators (ISOs).[43] And of the 55 GW in delayed clean power projects, wind projects are facing the longest delays, stretching out to 16 months.[44] An increase in local and state restrictions and contestations of renewable projects over the past year also impacted wind more than other sources (see the sidebar titled "Generative AI's impact on renewable deployment bottlenecks").[45]

Generative AI's impact on renewable deployment bottlenecks

While generative AI is being used to generate climate disinformation, fueling some of the opposition to renewables, it is also powering new tools for developers to assess community sentiment toward renewables and automate permitting and siting.[46] For the latter, generative AI can help select the best locations for renewable energy installations, considering wind patterns, solar exposure, and environmental impact. It can also suggest the best solar panel layout to maximize generation and design the most efficient blades with peak aerodynamics for wind. In 2024, more developers are expected to use generative AI tools to inform and accelerate renewable project decisions, processes, configurations, and community engagement.

Offshore wind faces high capital requirements, long project development and permitting timelines, and locked-in power sales contracts.[47] Developers executing agreements signed during low inflation now face higher financing costs and 40% jump in equipment and construction costs over the past year, while state policymakers may not be willing to provide additional support that could increase consumer costs.[48] Four contract renegotiations and three cancellations have imperiled half the of the offshore wind pipeline.[49] Federal and state action to expedite permitting, ease financing, and adjust incentives may be needed to keep projects in the pipeline and on track for meeting targets.

In 2024, the tide is expected to start turning as the industry gets more steel in the water and adapts to the new seascape. Developments to watch include the start of operations at Vineyard Wind in Massachusetts;[50] the construction of the country's largest offshore wind project in Virginia;[51] the Oregon, Central Atlantic, and second Gulf of Mexico lease sales;[52] and the pursuit of coordinated procurement in three northeastern states.[53] Flexible structures and market participants are expected to reshape agreements. Following Treasury guidance, more developers may seek to improve project economics by siting onshore grid connections in the energy communities that line US coasts and qualify for an adder.[54] There may also be greater uptake of the adder from the growing pipeline of onshore wind-repowering projects since most are located in energy communities.[55]

Tackling Transmission

Transmission is a factor in most constraints on renewable deployment. Regarding the top cost constraint survey respondents identified (figure 2), capturing the full customer benefit of low-cost renewables hinges on transmission. Insufficient capacity drove up congestion costs by 72% in 2022 over the previous year to US$20.8 billion.[56] Interregional and regional transmission would need to more than double and quintuple, respectively, to meet high clean energy growth projections by 2035.[57]

IIJA and IRA programs and grants could start tackling transmission issues in 2024. These include the DOE's announced plans to accelerate high-voltage transmission line permitting,[58] US$3.9 billion in grants from the Grid Resilience and Innovation Partnerships Program,[59] and US$1.3 billion in grants for three interregional grid projects.[60] At the beginning of the year, we'll be watching for the DOE's release of additional Transmission Facilitation Program funding and US Federal Energy Regulatory Commission interconnection rule compliance plans, as well as complementary ISO initiatives to reduce interconnection queues.[61] We expect to see more corporations participate in Federal

Energy Regulatory Commission regulatory filings as transmission constraints jeopardize their renewable targets.[62] At the same time, IRA and IIJA boosts to renewable development could significantly exacerbate pressure on transmission bottlenecks in 2024.

2. Reshoring Clean Energy: Supply chains shorten and strengthen

A domestic clean energy manufacturing revival is underway as producers restore to better capitalize on IRA tax credits and meet demand from renewable developers chasing domestic content adders. Since the IRA passed, companies have announced US $91 billion of investments in over 200 manufacturing projects, including US $9.6 billion in 38 solar projects, US $14.4 billion in 27 storage projects, US $1.4 billion in 14 wind projects, and US S$54 million in six hydrogen projects, closely tracking investment levels in their respective renewable energy sources.[63] These projects' shortened supply chains could increase transparency and resilience while decreasing emissions and exposure to geopolitical vicissitudes.

Solar and storage set to surge downstream and start extending upstream.

Announced projects could more than triple this year's solar photovoltaic module capacity in 2024, grow it by an order of magnitude by 2026, and meet US demand before 2030 (figure 3)[64]—a striking reversal from US import dependence for 85% of supply in 2022.[65] While China currently produces 83% of the cells and polysilicon and 97% of the wafers that go into modules,[66] new domestic polysilicon capacity and the United States' first cell, wafer, and ingot manufacturing plants are slated to come online in 2024.[67] This reshoring is premised on balancing domestic panels' 40% price premium[68] and the 40% tax incentive[69] for manufacturers that use 40% domestic components.[70] The latter part of this equation may be at risk because 40 GW of module capacity planned by 2025 may not qualify for a domestic content adder due to a lack of sufficient cell capacity.[71] Meanwhile, solar imports more than doubled in the first eight months of 2023 amid global overcapacity that drove prices to record lows, placing half the pipeline at risk of delays or cancellation.[72] In 2024, the enforcement of trade rules and the Uyghur Forced Labor Prevention Act and the expiration of waivers on duties covering solar cell and module imports from Southeast Asia in June could address overcapacity concerns. And more far-reaching final Treasury guidance on the domestic content adder could encourage upstream investment in wafers, ingots, and polysilicon.

Storage is on a similar trajectory to solar. Announced projects could drive almost eightfold growth in battery manufacturing capacity in 2024. Planned cell production could grow the US global capacity share from 4% in 2022 to 15% by the end of the decade in a segment where China currently holds a 79% share.[75] Yet, lithium-ion battery imports also reached a record high in 2023[76], and the US continues to be fully dependent on imports for some upstream supply chain components.[77] Companies have announced projects to manufacture copper foils and cathode active materials, 100 GWh for each, but none are expected to come online in 2024.[78] Expanded Uyghur Forced Labor Prevention Act enforcement on batteries could provide further impetus to domestic supply chain development in 2024.

Wind Homes in On Offshore Gap

The wind supply chain is more domestically rooted and evenly distributed across components than solar and storage, but the little capacity change planned for 2024 may be raising concerns offshore. China leads the global market for wind, too: With turbines selling at prices 70% lower than their Western counterparts, Chinese manufacturer exports jumped to 70% of wind turbine orders announced in the first half of 2023.[79] The greater role of developer relationships, specifications, and incentives in offshore wind has relatively insulated its supply chain from import pressure. On the other hand, infrastructure development and capacity additions have lagged demand from offshore wind developers seeking access to domestic content adders that could improve project economics. Meeting the Biden administration's offshore wind target could require a US$22.4 billion investment in 34 additional manufacturing facilities, 10 dedicated vessels, and 10 ports.[80] Whether projects proceed in 2024 hinges on the offshore wind pipeline's solidity. The development of 18 planned component manufacturing facilities[81] and the collaboration between nine East Coast states and federal agencies on offshore wind supply chain buildout are anticipated.[82]

Hydrogen Electrolyzers Take Root

Electrolytic hydrogen is embryonic at a time when the industrial policy is ascendant. The global production landscape is still in flux, providing a major opportunity to restore, and the global production landscape is still in flux, providing a major opportunity to restore renewables. The high-efficiency electrolyzers that dominate the US pipeline are more economical despite higher upfront costs, but they are exposed to competition from low-cost manufacturing regions.[83] China is already expected to account for over half of global capacity in 2023.[84] In the United States, companies have announced 9 GW of electrolyzer manufacturing capacity, under a quarter of which is expected to come online in 2024.[85] Most projects remain at early stages as the market awaits Treasury guidance on green hydrogen that could trigger a burst of electrolyzer demand.[86] The Department of Energy (DOE) estimates electrolyzer capacity would need to grow at a 20% compound annual growth rate to meet demand through 2050.[87] In 2024, we will be watching the relative competitiveness of manufacturers focused on scale versus modular approaches and their ability to keep pace with demand. Gaps could prompt domestic content requirements for electrolyzers.

Critical Mineral Crimp Tightens

The IRA has driven up energy transition demand for the critical minerals that underpin renewable supply chains. By 2035, this demand is expected to rise 15% and 13% higher than pre-IRA numbers for lithium and cobalt, respectively, which are needed for storage; 14% for nickel, which is in storage, wind, and hydrogen supply chains; and 12% for the copper needed across all energy transition technologies.[88] Meanwhile, domestic and free trade agreement country supply that could qualify for IRA incentives is limited. China refines around half of global copper production, two-thirds of lithium, three-quarters of copper, and four-fifths of nickel.[89] And Indonesian nickel, which accounts for half of global mining capacity, is mostly Chinese invested.[20] Underinvestment in mining amid currently low prices and long lead times for new projects stretching over a decade could yield yawning supply gaps. Shortages ranging from 10% to 40% across these minerals are expected by 2030.[21] In 2024, the impact of China's graphite export controls on critical mineral projects should be observed. The beginning of massive shifts in the lithium market from both the supply and demand sides may also become apparent. The discovery of the world's largest known lithium deposit in Nevada at the end of 2023 is a potentially meaningful change. The development of lithium alternatives, such as sodium storage batteries, could accelerate as manufacturers use generative AI to develop new molecules for testing.

Trends to watch as renewable energy companies restored in 2024 include the following:

- Companies are pursuing **strategic reshoring joint ventures** to secure a stake in the emerging domestic supply chain. For example, one of the largest renewable developers holds majority ownership and agreement to offtake 40% of the output from a new solar panel plant jointly developing with a solar manufacturer.[24] And a major solar manufacturer became the largest shareholder of a US polysilicon manufacturer, striking a 10-year take-or-pay agreement that helped restart the plant's production.[25] Critical mineral mining projects are also seeing direct investments from customers.[26]

- **Supply chain digitalization** is helping companies increase transparency, efficiency, and awareness of competitor demand. It can enable monitoring of environmental, social, and governance practices and compliance with labor and US and free trade agreement content requirements under the IRA and the Uyghur Forced Labor Prevention Act.

- Clean energy manufacturers are developing **end-of-life management and recycling** of solar panels, wind blades, batteries, and electrolyzers to reduce waste and recover critical minerals.[97] Battery-metal recycling startups raised record funding in 2022.[98] Since the IRA passed, six companies have announced investments in battery and wind blade recycling.[99] Two have received Loans Program Office conditional loan commitments. These projects could help address critical mineral shortages.

3. Reskilling the workforce: Unlocking the talent bottleneck is key to decarbonization

Sustaining a record buildout of renewables and domestic supply chain will require growing and (re)training a workforce with the right skills in the right places. Over the past two years, clean energy jobs have grown 10% faster than overall US employment.[100] There are currently 3.3 million clean energy jobs, the majority of which are in energy efficiency (68%), followed by renewable generation (16%), clean vehicles (11%), and storage and grid (5%).[101] Looking ahead, wind turbine service technicians and solar photovoltaic installers bookend the projected 15 fastest-growing occupations from 2022 through 2032, when IRA provisions are scheduled to sunset, with 45% and 22% growth, respectively.[102]

The IRA, the IIJA, and the Creating Helpful Incentives to Produce Semiconductors Act are expected to drive massive job creation over their lifetime: 19 million job years, or around 3 million jobs per year.[103] Compared to jobs in the overall workforce, a higher share—more than two-thirds—of direct jobs are available to workers without a bachelor's degree. The direct jobs created offer higher-median wages on average, but benefits and unionization rates are lower, and women and other minority groups are underrepresented, according to current data.[104] Announced manufacturing and generation projects across solar, storage, wind, and clean hydrogen plants, and their supply chains anticipate the creation of 72,557 annual construction jobs over five years and 24,193 annual operations and maintenance jobs over the lifetime of the plants (figure 4). However, the current half-a-million workforce shortage in the construction sector could constrain the buildout.[105] And while US green job postings grew 20% in 2022, green talent only grew 8.4%, revealing a growing skills gap.[106] Workforce challenges may be greatest in new industries such as clean hydrogen. The seven hydrogen hubs, expected to create 324,280 direct jobs across 16 states, include multiyear workforce plans to address these challenges.[107] More immediately though, the 3.6 GW of currently funded electrolyzer capacity is slated to start coming online in 2024.[108] The new facilities are expected to create around 12,000 infrastructure development jobs over the next two years, as well as 1,600 permanent operational jobs.[109] Deloitte analysis identified skill gaps across core roles and found that there is already a shortage of workers with the required skills, including electrical engineering, manufacturing processes, computer science, tooling, mechanical engineering, automation, and machining.[110] And the training pipeline for the relevant workforce of welders, machinists, and engineers across higher education institutions, including two- and four-year colleges, technical, and trade schools, shows potential talent supply challenges in three of the host states.

Generative AI is also reshaping renewable workforce needs. The solar and wind electric power generation industry includes five of the top 10 most AI-intensive occupations—occupations with the largest share of job postings demanding AI skills.[111] The most significant of these occupations in the industry are engineering professionals. Talent acquisition for these roles is already challenging, given the competition with other sectors.[112] In the administrative segment, generative AI facilitates the submission of daily work orders, supply part requests, asset maintenance, and bidding.[113] The core construction and maintenance workforce segments have the lowest penetration potential, 6% and 4%, respectively.[114] Here, AI is enabling full automation of the sector's most arduous work, such as offshore wind turbine inspections, which AI-powered drones could perform on spinning turbines at a lower cost.[115] At a broader electric and gas utility level, while most occupations have generative AI exposure, five occupations, including core electrical powerline installers and repairers, have none.[116]

It is crucial that the core power sector, including power, utilities, and renewable developers, align their decarbonization and workforce planning to ensure the sector is able to continue decarbonizing and other sectors such as transportation, buildings, and manufacturing (figure 5). These sectors are helping address some of the bottlenecks constraining the core power sector. For example, transportation electrification, clean fuels, and energy-efficient buildings can help address grid bottlenecks, while domestic clean energy manufacturing should help address supply chain challenges. However, the workforce bottleneck encompasses both the core and broader sectors.

More developers are expected to implement strategies with the following four following elements to help unlock the workforce bottleneck in 2024:

- **Equity:** Upskill the existing workforce in energy communities and help create new, diverse talent pipelines from local, untapped labor pools alongside workplace accommodations to address spatial, identity, and family structure inequities.

- **Quality:** Create purposeful, high-wage, credentialed jobs with portable skills and clear upwardly mobile career paths.

- **Agility:** Continuously assess skills gaps and training timelines (in weeks, months, and years), and proactively align with decarbonization strategy and technology timelines, the pace of domestic supply chain development, and the pace of digitalization, including the use of AI in the sector.

- **Comity:** Collaborate with ecosystem partners, including educational institutions, trade schools, high school career academies, and technical training groups; local, state, and federal governments; unions; industry associations; and philanthropic organizations to help develop registered apprenticeships, courses, and small business training support.

In 2024, the US Treasury's final guidance on prevailing wage and apprenticeship requirements should take hold and help facilitate a pipeline of renewable energy apprentices that could alleviate shortages. Also anticipated is greater renewable developer and utility uptake of funding from the 54 IIJA and IRA programs that can be deployed for green workforce development.[117]

4. Renewables as a resilience strategy: Amid widespread misperceptions, renewables can save the day

As the frequency and intensity of extreme weather events, outages, and potential electricity supply shortages rise, renewables have often outshined conventional power sources, generating electricity when the latter could not. Renewables are increasingly becoming a resilience strategy, especially when coupled with storage. This reality does not match with perception, however. More respondents of the Deloitte 2023 power and utilities industry survey were concerned about the resilience of renewables than supply chain challenges and the interconnection queue. Most survey respondents believe that gas, followed by nuclear power, is the most resilient to extreme weather events in their territory, while renewables ranked the lowest (figure 6).

But, in contrast to the surveyed respondents' perception, experience with a record number of extreme weather events and outages over the past year shows that gas poses greater reliability challenges than renewables. For instance, during Winter Storm Elliott, unplanned generation outages reached a record 90.5 GW across the Eastern Interconnection, mainly driven by natural gas infrastructure reliability issues.[118] The impact on the Pennsylvania-New Jersey-Maryland Interconnection (PJM) was especially striking given the grid's historical reliability, familiarity with cold weather, and location atop shale gas basins that directly supply many power plants. A fifth of gas plants, including new ones, failed to ramp up to half capacity during the grid's two emergency calls due to a range of malfunctions across the system, from mechanical problems to frozen transmitters, valves and wells, pipe pressure issues, compressor stations failures, and supply scheduling gaps.[119] Gas accounts for 46% of PJM capacity, but drove 70% of forced outages.[120] Shifting seasons and states, in the summer, thermal plant outages unexpectedly went above the 11,000 MW red line, which, according to the Electric Reliability Council of Texas (ERCOT), could put its grid at risk.[121]

Nuclear also faces increasing reliability concerns as warmer and lower water levels caused by climate change impact operations. Over the past summer—a season when nuclear is most needed to meet power demand—a hot weather alert factored into a shutdown of the nuclear Vogtle plant reactor in July.[122] Another nuclear plant shut down later in the summer due to coolant leakage, contributing to a total of 31 unplanned nuclear outages from January through October 2023 and a 25% rise in total nuclear capacity outages in the summer of 2023 versus that of 2022.[123]

Meanwhile, renewables paired with storage are taking on the role of gas peakers that can quickly respond to demand spikes and avoid blackouts. During Winter Storm Elliott, strong wind generation helped the Midcontinent Independent System Operator meet demand and continue exports despite 49 GW of forced outages.[124] When Texas experienced 10 demand records this summer, batteries discharging in the evening played a key role in avoiding blackouts, while solar and wind generation covered more than a third of demand load in ERCOT during the day and helped prevent power price spikes.[125] As a result, ERCOT included storage for the first time as a resource able to meet high net load in its fall Seasonal Assessment of Resource Adequacy for fall 2023.[126] Similarly, renewables contributed to a fifth of generation during a heatwave that drove record loads in the Southwest Power Pool.[127]

On the distributed renewable front, when the California Independent System Operator called for electricity conservation on August 17, an aggregation of 2,500 residential storage systems were activated for the first time to deliver 16.5 MW of solar power to the grid.[128] Some utilities are subsidizing residential battery installations to create such AI-orchestrated aggregations to draw on during peak demand.[129] In the aftermath of winter storms and flooding, the Vermont Public Utility Commission lifted caps on programs supporting residential storage the utility can tap during emergencies.[130] Meanwhile, generative AI is enabling greater solar photovoltaic module power forecasting and proactive mitigation of extreme weather and cyberattacks.[131] In its latest forecast, the North American Reliability Corporation not only warned about an elevated risk of blackouts across the country this winter,[132] but also showed that some states rapidly transitioning to renewables are among those at lowest risk of outages.[133] The year 2024 may be when perception catches up with reality.

5. Renewable technology, redefined: Underground renewables could resurge

Technologies expected to become more apparent over the next year are transforming renewable capabilities, synergies, and deployment potential. Renewable deployment over the past decade has primarily told a success story for onshore wind, solar, and storage growth amid dramatically falling cost curves. On the other side, the intermittency and the geographical land use and industrial end-use limitations of renewables are often cited as reasons these resources cannot replace gas as a direct and back-up fuel that can be deployed anywhere and tapped anytime. Yet, renewables that can do just that while supporting grid resilience have made strides this year in moving technological innovation toward commercialization. Two of these renewables are longstanding but often overlooked underground resources: geothermal and renewable natural gas (RNG).

Enhanced geothermal systems (EGS) have expanded the potential to capture the heat of the earth. While the United States is the global leader in geothermal electricity production, geothermal only accounts for 0.4% of US utility-scale generation and is concentrated in Western states, with natural hot water reservoirs in permeable rock at low depths.[134] EGS could push this share past 6% by 2035, the target date for a 90% cost reduction in EGS, to US$45/MWh under the DOE's Enhanced Geothermal Earthshot initiative.[135] EGS uses technology from the oil and gas industry to help create artificial reservoirs and access the omnipresent heat available below the earth's surface. Cost reductions are achieved from advanced sensing and drilling, which accounts for half the cost of geothermal projects.[136] Developer use of generative AI to assess seismic data and guide drilling has further driven down costs.[137] EGS can also bring more value by using reservoirs for long-duration storage and direct air capture. This year saw breakthrough announcements from the DOE-funded Frontier Observatory for Research in Geothermal Energy and geothermal energy startups on demonstrating commercial viability and breaking ground on the world's largest EGS plant. While geothermal accounts for less than 3% of power purchase agreements (PPAs), the megawatt (MW) volume of geothermal PPAs quintupled from 2021, when the world's first corporate geothermal PPA was signed, to 2022; next-gen geothermal accounted for more than half of the MW volume in 2023.[138] The EGS project pipeline and PPA market are likely to continue strengthening in 2024 to meet demand from the growing number of corporations with 24/7 decarbonization targets.

RNG project development has accelerated recently as well. RNG only accounts for 0.5% of the gas market and mostly serves the transportation sector but could grow tenfold by 2050 as usage expands to power and heat.[139] Demand has overtaken supply as developers seek to begin construction on RNG facilities before 2025 to qualify for the IRA Section 48–qualified biogas property investment tax credit. The number of operational plants jumped from 230 in 2021 to 300 through the end of July 2023, while the pipeline of plants planned or under construction grew from 172 to 481 for the same period.[140] Landfill and livestock operations are driving most of RNG capacity growth anticipated to come online in 2024.[141] These two sources of feedstock account for more than half of US methane emissions, which RNG production can prevent from being vented.[142] Food waste and wastewater treatment projects are also growing. New feedstock streams that could start expanding in 2024 may include forest waste from vegetation management related to wildfire prevention. The use of RNG as a feedstock to produce biohydrogen and sustainable aviation fuels could also take off depending on Treasury guidance on the hydrogen tax credit and carbon accounting. Two major US hydrogen projects are currently using RNG to produce hydrogen.[143] Also important to consider is US Environmental Protection Agency rulemaking that could grant renewable credits for RNG used in plants generating power to charge electric vehicles. States with supportive policies could further increase RNG demand (figure 7).

In 2024, renewable developers may consider expanding into renewable resources resurging with innovative technologies. Geothermal and RNG can help developers diversify their renewable portfolios and capitalize on new synergies between intermittent and baseload renewables, and between electrons and molecules.

What to expect in 2024

In 2024, the renewable energy industry could expect to see the historic climate legislation take greater effect as tax credit guidance is finalized, more Loans Program Office loans are issued, and more programs release IRA grant funding, only 10% of which has been disbursed thus far.[144] The massive public and private investment and channeling of capital toward the clean energy transition could propel solar and storage deployments to continue soaring, onshore wind to recover, and residential technologies to pick up speed. Offshore wind and green hydrogen industries could establish a foothold, while underdeveloped renewables could play a greater role in clean energy portfolios. Meanwhile, a clean energy–driven manufacturing renaissance could provide opportunities to develop more resilient renewable supply chains across the country. The surge in renewable projects and domestic manufacturing also calls for a bigger and smarter grid, a skilled workforce to build and operate the plants, and a smooth process to develop both. Challenges in these areas should be addressed in 2024 to help keep the country and corporations on track to achieve their climate goals.

Signposts to watch include the US Treasury's guidance on hydrogen and domestic content adders, the impact of IRA and IIJA funds on workforce development, and Federal Energy Regulatory

Commission and DOE actions on grid reform and buildout. This El Niño year may bring more extreme weather events that could call on renewable resources to support the grid. Finally, a quickly expanding range of use cases may grow generative AI's foothold in renewable operations, workforce planning, and distributed aggregations supporting resilience.

The world's largest experimental tokamak nuclear fusion reactor is up and running.

Located north of Tokyo, the six-story high JT-60SA could spur advancements towards the 'Holy Grail' of renewable energy.

By <u>Andrew Paul</u>

Posted on Dec 4, 2023, 12:00 PM EST.

A view of the assembled experimental JT-60SA tokamak nuclear fusion facility outside Tokyo, Japan. JT-60SA.org

Japan and the European Union have officially inaugurated testing at the world's largest experimental nuclear fusion plant. Located roughly 85 miles north of Tokyo, the six-story, JT-60SA "tokamak" facility heats plasma to 200 million degrees Celsius (around 360 million Fahrenheit) within its circular, magnetically insulated reactor. Although JT-60SA first powered up during a test run back in October, the partner governments' December 1 announcement marks the official start of operations at the world's biggest fusion center, reaffirming a "long-standing cooperation in the field of fusion energy."

The tokamak—an acronym of the Russian-language designation of "toroidal chamber with magnetic coils"—has led researchers' push towards achieving the "Holy Grail" of sustainable green energy production for decades. Often described as a large hollow donut, a tokamak is filled with gaseous hydrogen fuel that is then spun at immense high speeds using powerful magnetic coil encasements. When all goes as planned, intense force ionizes atoms to form helium plasma, much like how the sun produces its energy.

Speaking at the inauguration event, EU energy commissioner Kadri Simson referred to the JT-60SA as "the most advanced tokamak in the world," representing "a milestone for fusion history."

"Fusion has the potential to become a key component for energy mix in the second half of this century," she continued.

But even if such a revolutionary milestone is crossed, it likely won't be at JT-60SA. Along with its still-in-construction sibling, the International Thermonuclear Experimental Reactor (ITER) in Europe, the projects are intended solely to demonstrate scalable fusion's feasibility. Current hopes estimate ITER's operational start for some time in 2025, although the undertaking has been fraught with financial, logistical, and construction issues since its groundbreaking back in 2011.

Experts alongside Simson believe creating sustainable nuclear fusion would mark a revolutionary moment that could ensure an emission-less, renewable energy future. Making the power source a feasible reality, however, is fraught with technological and economic hurdles. Researchers have chased this goal for a long time: The world's first experimental tokamak was built back in 1958 by the USSR.

While researchers can now generate fusion energy at multiple facilities around the world, it is usually at a net loss. By advancing the technology further at facilities like JT-60SA, however, industry experts think that it is only a matter of time until fusion reactors regularly achieve net energy production gains.

In the meantime, another possible road to fusion energy is making its own promising gains. Earlier this year, the National Ignition Facility (NIF) at Northern California's Lawrence Livermore National Laboratory achieved a net energy gain for the second time using what's the inertial confinement fusion method. In this process, a high-powered laser is split into 192 beams that then hit a capsule containing a pellet of tritium and deuterium. The resultant X-rays generate pressure and temperatures that then initiate fusion.

No matter which process—be it tokamak reactors or ICF lasers— a successful nuclear fusion facility could play a significant role in finally shifting humanity away from fossil fuels.

This post on Exxon/Mobil is in this book because of the extraordinary performance of this company today (2024) to take the fossil fuel industry to new levels to improve its delivery of fuels that will undoubtedly bridge the fuel gap and encourage other companies to do the same in support of industries that need badly needed breathing room to adjust to requirements being suggested by the NGD, such as the transportation and airline companies struggling with what to do for fuels. The battery knights want to have trucks to use effectively killing the long-haul trucking industry.

I am not indifferent to the existence of electric or other fuel solutions, but the downside of electric is really bothersome to me. I do not see a favorable outcome due to the production, salvage, and disposal of defunct batteries. Yes, I am aware of parts salvage, but how about the really dangerous stuff? Reminds me of California heralding that they were going to tackle the disposal of used solar panels (not just a few, but millions of them, which are carcinogen-loaded), and at the end of the meeting, they had no solution, and the landfills got them.

Exxon Mobil buys Pioneer Natural Resources and rolls back Pioneers clean oil from 2050 to 2030.

Exxon Mobil Advancing Climate Solutions Executive Summary Getting the planet on a path to net zero requires unprecedented innovation and collaboration at immense scale. The ongoing societal effort is critical but must avoid economic hardships and market disruptions that result from energy and product shortages. Solving this challenge is not an "either/or" proposition. It's an "and" equation. One that requires an increase in energy supply and reduction in greenhouse gas emissions – improved energy security and thoughtful progress in the energy transition. Given the skills and capabilities required, there's no question that the energy industry plays a critical role – on both sides of this equation. ExxonMobil is doing our part. Since 2016, we've significantly reduced our Scope 1 and 2 operated emissions. And we've got plans to do more. | 2024 Advancing Climate Solutions | Executive Summary 3 People Technology Scale Integration Functional excellence Corpus Christi Chemical Complex 2050 net-zero ambition (cont'd) Our net-zero ambition is backed by a comprehensive approach centered on detailed emission-reduction roadmaps. We completed these roadmaps in 2022 and continue to update them to reflect technology and policy, and to account for the many potential pathways, and the pace of the energy transition. We are using this approach in our Permian Basin unconventional operations, where we are on track to achieve our industry-leading plans to reach net-zero Scope 1 and 2 emissions by 2030. Beyond reducing emissions in our own operations, we see the opportunity to use our core capabilities to help other essential industries and customers lower their emissions. This is an immense opportunity with an addressable Competitive advantage The same competitive advantages that have underpinned the success of our traditional businesses for more than 140 years are the foundation of this world scale Low Carbon Solutions business.

market potentially measured in the trillions of dollars by 2050.[4] That's why we established ExxonMobil's Low Carbon Solutions business. We're working to profitably grow a leading position in these new emission-reduction markets, with a focus on the global economy's hard to-decarbonize sectors – like heavy industry, power generation, and commercial transportation. These are critical sectors where cost-effective solutions are lacking and where we can make a unique, significant, and lasting contribution. | 2024 Advancing Climate Solutions | Executive Summary 4 Corpus Christi Chemical Complex The challenge is enormous. To tackle it, the world needs industrial-scale solutions. We need them deployed globally and at a much lower cost than today. That will require continued advances in technology, and clear and consistent government policies that catalyze investments in the near term. Additionally, the world will need to establish a new industry – a carbon-reduction industry – and a market that pays for the cost of emission reductions. The skills and capabilities required to address these complicated challenges play to ExxonMobil's strengths and align with our strategic priorities: Leading performance Industry leader in operating and financial performance. • Essential partner Value through win-win solutions for our customers, partners, and broader stakeholders. • Advantaged portfolio of assets and products outperform competition and grow value in a lower-emission future. Innovative solutions, new products, technologies, and approaches to accelerate large-scale deployment of solutions essential to modern life and lower emissions. • Meaningful development Diverse and engaged organization with unrivaled opportunities for personal and professional growth doing impactful work to meet society's needs. 4 | 2024 Advancing Climate Solutions | Executive Summary 5 About this report This year's edition of ExxonMobil's Advancing Climate Solutions Report describes our resolve to drive meaningful change, the results we're already delivering, and the resiliency of our plans under a wide range of future scenarios. This Executive Summary highlights the noteworthy progress we continue to make toward achieving our 2030 emission-reduction plans and our 2050 net-zero ambition. • Reducing methane emissions. • Building our Low Carbon Solutions business.

We encourage you to visit our website to explore greater detail on these topics and others related to our actions to address the risks of climate change across our businesses. Making progress Over our history and across the globe, we have built industries where none existed before. We see this today with our developments in Papua New Guinea and Guyana. At our core, we're a technology company that uses our science and engineering capabilities to bring value-added solutions to partners and customers. We do this in a variety of ways using unique advantages in scaling technology and delivering complex, large-scale projects safely, reliably, and at an industry-advantaged cost. We're developing molecules that cost-effectively meet the ever-evolving needs of society. We're unlocking critical oil and natural gas resources trapped in geologic formations around the world. And we're capturing and safely storing emissions for hard-to-decarbonize industrial processes. Of course, our past successes and current strengths stem from the commitment, experience, and capabilities of our people. Their skills, tenacity, and resiliency are the bedrock on which our company is built. If you were to make a list of the biggest challenges facing humankind right now, addressing poverty and climate change would be at the top. At the same time, if you were to make a list of the companies that have a credible chance of improving access to affordable energy and other products that are critical to improved living standards and reducing emissions, ExxonMobil would also be at the top. The strategy we've developed, the organization we've built, and the businesses we're focused on position us to grow and create value for many decades to come, regardless of the pace of the transition. At our core, we're a technology company that uses our science and engineering capabilities to bring value-added solutions to partners and customers. Guyana Prosperity FPSO | 2024 Advancing Climate Solutions | | 2024 Advancing Climate Solutions | Executive Summary Executive Summary 6 Making real progress toward solving the "and" equation ExxonMobil is delivering both sides of the "and" equation – meeting society's needs for energy and essential products and reducing emissions. • We achieved record production from our projects in the Permian Basin and Guyana in the second quarter of

221

2023, up more than 20% from a year earlier.6 • We added 250,000 barrels per day of refining capacity in early 2023 in Beaumont, Texas. The extra supply helps reduce rising price pressures, easing the impact on consumers and businesses. It was the largest refinery expansion in the U.S. since 2012.7 • We started up a chemical expansion project at Baytown, Texas, that has capacity to deliver 750,000 tons per year of products that are used by manufacturers to make stronger and lighter auto parts, construction materials, packaging, and more.8 • We've cut operated methane emissions in half since 2016, eliminated all of our high-bleed pneumatic devices in U.S. operated unconventional production, and established our Center for Operations and Methane Emissions Tracking (COMET). When fully deployed, COMET is expected to provide around-the-clock remote monitoring capabilities in the region. • We eliminated routine flaring in our Permian Basin operated assets, in line with the World Bank's Zero Routine Flaring Initiative,9 which is a key part of our 2030 goal of achieving net-zero Scope 1 and 2 greenhouse gas emissions from our unconventional operated assets in the Permian. • We electrified our drilling fleet in the Permian Basin and deployed our first electric fracturing units to further reduce emissions intensity.10 • We acquired Denbury Inc., which expands our Low Carbon Solutions business opportunities by leveraging the largest CO_2 pipeline network in the United States.11 • We signed landmark CO_2 offtake agreements with a major fertilizer producer, a steel manufacturer, and an industrial gas company to capture, transport, and store up to 5 million metric tons of CO_2 per year. That's equivalent to replacing approximately 2 million gasoline-powered cars with electric vehicles,12 which is roughly equal to the total number of EVs on U.S. roads today.13,14,15 • We began drilling for lithium in southwestern Arkansas – a process that holds great promise to address the growing needs of the EV battery markets. Increasing energy and product supply & Reducing greenhouse gas emissions5 | 2024 Advancing Climate Solutions | Executive Summary 7 2030 greenhouse gas emission-reduction plans16,17 Since 2016, we've reduced our operated greenhouse gas emissions intensity by more than 10%, and our 2030 plans are expected to drive further reductions. Corporate-wide greenhouse gas intensity 2030 plan: 20-30% Corporate-wide methane intensity 2030 plan: 70-80% Upstream greenhouse gas intensity 2030 plan: 40-50% Corporate-wide flaring intensity 2030 plan: 60-70% Our plans to reduce emissions through 2030 include: Achieving net-zero Scope 1 and 2 greenhouse gas emissions in our Permian Basin unconventional operated assets. • Deploying carbon capture and storage, hydrogen, and lower-emission fuels in our operations. • Further reducing methane emissions at operated assets in alignment with the Global Methane Pledge and with Aiming for Zero Methane Emissions, developed by the Oil and Gas Climate Initiative. • Further reducing flaring in upstream operations to meet the World Bank Zero Routine Flaring Initiative. • Integrating lower greenhouse gas energy sources into our facilities through long-term power purchase agreements and electrification. • Improving energy efficiency in our businesses by evolving operational and maintenance processes. • Substituting low-carbon hydrogen for natural gas to reduce emissions from furnaces. • Deploying innovative solutions to further reduce greenhouse gas emissions with future advancements in technology and supportive policies. Did you know? ExxonMobil is a leading purchaser of renewable power.18 view web module | 2024 Advancing Climate Solutions | Executive Summary 8 >10% reduction in corporate-wide greenhouse gas (GHG) emissions intensity19 Corporate-wide operated GHG emissions intensity (T CO_2e/100 T) 2022 year-end actual: Versus 2016 levels. Applies to Scope 1 and 2 GHG emissions from operated assets. 0 10 20 30 2016 2022 2030 plan 0.00 0.02 0.04 0.06 0.08 2016 2022 2030 plan 0 10 20 30 2016 2022 2030 plan NDCs 0 5 10 15 2016 2022 2030 plan Corporate-wide operated methane emissions intensity (T CH_4/100 T) 2022 year-end actual: Upstream operated GHG emissions intensity (T CO_2e/100 T) 2022 year-end actual: Corporate-wide operated hydrocarbon flaring intensity (m3/T) 2022 year-end actual: Progress through year-end 2022 • Methane and flaring intensity reductions make up the bulk of our improvement. • Our actions to reduce emissions intensity significantly offset our growth. • Divestments did not meaningfully contribute to our intensity reductions. Others include power-purchase agreements, energy attribute certificates, and other changes. 2024 Advancing Climate Solutions | Executive Summary 9 Approach to reducing emissions in business planning We incorporate actions needed to advance our 2030 emission reduction objectives into our medium-term business plans, which we update annually. The reference case for planning beyond 2030, including impairment assessments and future planned development activities, is based on our Global Outlook. The Outlook considers the existing global policy environment, announced policy changes, advances, consumer preferences, and the historical precedents for each of these areas. It does not attempt to project the degree of future policy, technology advancement, or deployment necessary for the world or ExxonMobil to meet net zero by 2050. As additional policies are implemented and technology advances beyond our estimates, we incorporate those changes into the Outlook and update our business plans accordingly as part of our annual planning cycle. Potential GHG abatement options for ExxonMobil operated assets supporting 2030 GHG emission-reduction plans20 Roadmap T CO2e/100 T 2016 intensity 2022 intensity Energy efficiency Flare & methane minimization Operations/ reconfigurations Electrification/ PPAs/RECs/ high-quality offsets CCS, hydrogen, and/or future advancements 2030 intensity Positioning for a lower-emission future We have evolved our operating model, enabling efficiencies that better leverage the scale of an increasingly integrated company. At the same time, we have centralized many of the skills and capabilities required by our business, allowing us to improve allocation of critical resources; drive continuous improvement, including detection and measurement of emissions; and grow value. This serves us well in a variety of future scenarios, irrespective of the pace of the energy transition. Higher-cost options reflect the need for additional policy and continued advocacy. Abatement Cost Abatement cost Electrification Operations/ reconfigurations Energy efficiency Flare minimization Methane minimization CCS, hydrogen, and/or future advancements Electrification/ PPAs/RECs/ high-quality offsets CO2e mitigation Abatement curve Roadmap Abatement curve |

Reducing methane emissions Our plans to reduce methane intensity across our operated assets remain on track. These include reductions versus 2016 levels of 70%-80% in methane intensity and 60%-70% in flaring intensity by 2030. To get there, we're developing and deploying enhanced technologies from satellites to on-the-ground sensors for rapid detection and mitigation – starting with a focus on our highest methane emission sources. At the same time, we're continuing to develop and advocate for strong measurement and reporting frameworks to provide consistent, comparable, and most importantly, useful data to inform our methane mitigation efforts worldwide. In 2023, we took additional steps to further collaboration among government and industry partners, including deciding to join the United Nations Oil and Gas Methane Partnership 2.0. Our Permian operations make up about 16% of our total methane emissions. By rapidly advancing our plans in the basin, we're reducing emissions and developing solutions that we can refine and deploy in other parts of the world. As of year-end 2022, we have eliminated routine flaring in our Permian operations. With full deployment of our near-continuous monitoring program in the Permian by 2025, we expect our Center for Operations and Methane Emissions Tracking (COMET) to provide real-time monitoring of 700 sites across 1.8 million acres. Our progress in the Permian Basin guides our projects elsewhere. The pneumatic devices in our industry are, as a category, the largest source of routine methane emissions in our processes. That's why in 2020, we completed the elimination of high-bleed pneumatic devices across our U.S. unconventional production, and we're working to eliminate the rest by 2025. Through actions like these, we're eliminating potential sources of methane emissions while advancing our ability to detect and quantify others. We know we can't do it alone. Collaboration will be vital as we implement solutions to support society's net-zero future. By working with a wide range of universities, academic consortiums, environmental groups, and more, we're advancing leading-edge research and piloting innovative technologies to help the industry and our company measure, reduce, and report methane emissions.

view web module As of year-end 2022, we have eliminated routine flaring in our Permian operations. We're developing and deploying enhanced technologies from satellites to on-the-ground sensors for rapid detection and mitigation. 2024 Advancing Climate Solutions Executive Summary 11 Sustaining our commitment to R&D We determine which research projects to advance based on factors including advantage versus alternatives, ability to scale, alignment with core capabilities and key partners, and probability of commercial success. We employ thousands of scientists and engineers, including more than 1,500 Ph.D.'s. Their work drives our research in new materials, novel low energy processes, and improved means of CO_2 storage. Our scientists have written more than 1,000 peer-reviewed publications and received more than 10,000 patents over the past decade. In addition, we collaborate with more than 80 universities around the world, four energy centers, and several U.S. national laboratories. These collaborations have increased knowledge in key areas important to the energy transition, including fugitive methane emissions detection and modeling; optimization techniques to understand CO_2 storage; electrification of processes; lower emission fuels; and energy systems models. >1,000 peer-reviewed publications written by our scientists >1,500 Ph.D.'s employed >10,000 patents over the past decade >80 university collaborations around the world Shanghai Technology Center view web module 11 | 2024 Advancing Climate Solutions | Executive Summary 12 Investing in lower-emission solutions We're pursuing more than $20 billion in lower-emission investments from 2022 through 2027, in addition to the approximately $5 billion Denbury acquisition. About 50% of our lower-emission investments are targeted at reducing emissions from operated assets, with the balance going toward reducing the emissions of other companies. We're focused on customers in the heavy industry, power generation, and commercial transportation sectors. These sectors provide great economic value and generate significant emissions that aren't easy to cut. Together, these sectors account for about 80% of energy-related CO_2 emissions today. Carbon capture and storage, hydrogen, biofuels, and lithium align with our capabilities and have the potential to make a significant difference in these hard-to-decarbonize sectors. Denbury view web module 12 | 2024 Advancing Climate Solutions | Executive Summary 13 Carbon capture and storage The technology exists today to capture and store CO_2 from emission sources. Global agencies including the International Energy Agency, the U.N. Intergovernmental Panel on Climate Change, and the U.S. Department of Energy have concluded that permanent storage of CO_2 in appropriately selected geological formations is a safe and secure option.21 ExxonMobil has cumulatively captured more human-made CO_2 than any other company on the planet, and we're expanding our long-term storage capacity in anticipation of market developments. We have three of the largest third-party contracts to capture, transport, and store CO_2 – advancing projects that will help decarbonize a fertilizer company, an industrial gases company, and a steel company. The recent acquisition of Denbury expands our capabilities in this area. It provides ExxonMobil with the largest owned and operated network of CO_2 pipelines in the United States, including 900 miles of pipelines near the largest industrial complexes on the Gulf Coast. Combining Denbury's assets and our experience accelerates and expands our ability to help customers reduce their emissions. Ultimately, we see an opportunity to create a carbon capture and storage business with the capacity to reduce emissions across the Gulf Coast by more than 100 million metric tons per year.22 This transaction will help us do that at a lower cost and faster pace. Corpus Christi Baytown Beaumont Lake Charles Port Arthur Baton Rouge Houston TX LA MS AL Industrial emissions sources ExxonMobil industrial sites CO_2 storage sites Acquired CO_2 pipeline Enhanced oil recovery Note: All information shown is approximate (e.g., storage / pipeline location) and has potential to change as projects are developed and implemented. Denbury acquisition creates strong U.S. Gulf Coast CO2 infrastructure position | 2024 Advancing Climate Solutions | Executive Summary 14 Hydrogen We also have a long history with hydrogen, a zero-carbon energy source that can be used to reduce emissions in hard-to-decarbonize sectors including steel manufacturing, refining, and heavy-duty trucking, among others. In Baytown, Texas, we are developing the world's largest low-carbon hydrogen production facility. We are designing it to produce 1 billion cubic feet of hydrogen per day, using a process called "auto-thermal reforming "to separate the hydrogen and carbon atoms. We plan to use carbon capture and storage to sequester the CO_2 emissions. More than 98% of the associated CO_2 emissions produced by the facility – 7 million metric tons per year – are expected to be captured and stored. Biofuels We can also make a real difference with biofuels. Demand for energy-dense, lower-emission fuels is expected to grow rapidly, especially in the aviation, marine, and heavy-duty trucking industries. This growth creates opportunities to process biofuels and make drop-in replacements for today's fossil fuels. Our Product Solutions business is working to supply approximately 40,000 barrels per day of lower-emission fuel by 2025, with a further goal of 200,000 barrels per day by 2030. Lithium production is an exciting new business opportunity for us. We're working to apply our upstream and downstream expertise to recover and separate lithium from deep brine reservoirs. Using available technologies, we're working to produce this critical mineral more efficiently and with fewer environmental impacts than traditional hard rock mining – helping to grow a U.S.-based supply for the global battery and electric vehicle markets.

Author's notes:

Did someone say Hybrid cars? Of course, they did, and they are an interesting breed of automobile. The cars run on electric batteries and fuel, but did you know they are evolving? Lete review the reference materials I have that actually give us a brief on their current makeup and, oh my, tell us what are some of the issues with them.

Hybrid cars can have some environmental impacts, including:

Mining

The mining of rare-earth metals and other materials, such as lithium, copper, and nickel, for hybrid batteries and motors can release toxic compounds into the air and groundwater. Mining can also damage the environment by clearing areas of trees, grasses, and dirt to make room for digging and machinery.

Battery production

The manufacturing process for hybrid batteries is energy-intensive and uses a lot of raw materials, some of which are rare and hard to mine. The process also requires high heat and sterile conditions, and much of the energy used is not clean energy.

Air pollution

Every time a hybrid car is driven, it burns gasoline and emits exhaust into the atmosphere. While hybrids create less air pollution than gas-powered cars, even the most fuel-efficient hybrids still produce some air pollution.

However, hybrid cars can still have a lower overall carbon footprint than internal combustion engine vehicles over their lifetime. For example, a hybrid car emits 51.6 pounds of carbon dioxide every 100 miles, compared to 74.9 pounds for a conventional car.

The Hidden Environmental Impacts of Hybrid Cars

Mike Beaulieu '20, Managing Editor

December 20, 2019

A 2012 Toyota Prius.

As the global warming crisis continues to worsen, an emphasis has been put on switching to electric cars to protect the environment. But they may be just as harmful.

Vehicles that run on fossil fuels emit toxic gasses that get into the atmosphere, which causes many environmental activists to see hybrids and electric cars as the answer. Although these cars do not emit the same toxic gasses, the process in which they are made is also bad for the environment. Much of this problem lays in the production of the batteries and the mining of the metals necessary to make them.

The composition of the batteries of an electric car consists of rare metals like lithium and cobalt. Why does this matter? Because these metals take a tremendous amount of effort and machinery to acquire, contributing further to carbon emissions.

Not only does mining tax the environment, but the factory production of the batteries they are made into is even worse, according to *The International Council On Clean Transportation.*

"Lithium-ion battery production requires extracting and refining rare earth metals," said the ICCT, "and is energy intensive because of the high heat and sterile conditions involved."

Not only do the metals used to produce the batteries use up a considerable amount of energy, but much of the energy used is not clean energy.

"Most lithium-ion batteries in electric vehicles in Europe in 2016 were produced in Japan and South Korea, where approximately 25%–40% of electricity generation is from coal."

Many statistics, including those found on *Greentechmedia.com,* back up that large-scale emissions are being produced by the production of electric cars and lithium-ion batteries.

"2015, the latest year for which actual figures are available) were, on average, 20.4 percent too high for energy-related CO_2 emissions, and 42.7 percent too high for coal consumption."

From Northampton Mass, Senior Class president Nat Markey commutes to school in his Prius, speaking highly of its self-charging abilities.

"Our Prius recharges its own battery," said Nat, "so we don't have to plug it in and use energy from the grid."

Nat acknowledged the harmful carbon emissions the batteries release, but said he still believes the benefits outweigh the concerns.

"I've heard that there is also significant carbon emission from making hybrid and electric car batteries," said Nat, "But from what I've read, the increased fuel economy most likely outweighs this."

- cars
- hybrid

About the Contributor

Mike Beaulieu '20, Managing Editor

Mike is a four-year senior from Somers, Connecticut. He enjoys covering a lot of campus topics and campus news. His favorite person to interview is Franklin...

Hybrid Cars: Environmental Blessing or Ticking Time Bomb?

By

Jeremy Laue

December 9, 2022

I have a family member who scoffs every time they see a Toyota Prius, Nissan Leaf, or any car with a hybrid badge on it drive by.

Every time one went scooting by, they would say something like, "They think they are helping the environment, but the damage their batteries due to the environment is insane."

So, I started digging. I began by looking through polls on why people buy hybrid and electric cars, through white papers on hybrid car batteries' environmental impact, and at stories on why hybrid cars may not actually be helping the environment. I was searching to find out if hybrid batteries are indeed that big of a deal, and if there were perhaps other issues that weren't being talked about publicly.

This is what I was able to find; I'm not going to give you my opinion, just the facts. Armed with the information I'm about to give, you can then decide for yourself if hybrid cars are a blessing for the environment or a ticking time bomb.

Energy Source Impact

Plug-in hybrids and Battery Electrical Vehicles need to be plugged into a 120-volt power source in order to fully charge their electric only systems. The reason this can be problematic is that this means they are effectively tapping whatever type of power plant is producing that energy; be it coal, green energy, nuclear, etc. In the case of California, there is a minor impact on the environment due to the fact that 80% of their power is clean energy; solar, wind, hydroelectric, etc.

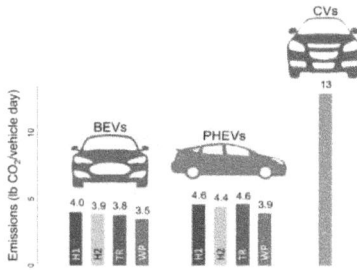

Figure 17. Comparison of total BEV and PHEV emissions with emissions from a conventional vehicle on a low carbon grid.

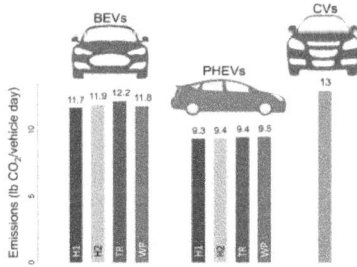

Figure 18. Comparison of total BEV and PHEV emissions with emissions from a conventional vehicle on a high carbon grid

Ev Emissions Impact report from National Renewable Energy Laboratory

This would be an example of a low carbon power grid environmental impact (see the top portion of the figure above). Now, if you have to charge your hybrid or electric vehicle in a state that gets most of its electricity from coal burning power plants, then you are ostensibly contributing to the problem of power plant emissions (an example of charging a vehicle in a high carbon energy grid can be seen in the bottom of the figure above). This shows the level of impact the source of charging energy can have on a vehicle's carbon emission.

Another factor is how much a person ends up driving on their gasoline engine in cases of a hybrid or plug-in hybrid.

It's important to remember that 45% of electricity in the United States is generated by coal power plants. This study from the Argonne National Laboratory shows that if a plug-in hybrid charges on an energy grid that uses mostly coal to generate electricity it would be responsible for emitting up to 10 percent more greenhouse gases than a conventional gas powered car, and up to 60% more than a standard hybrid.

Batteries

The batteries used in vehicles are not a new environmental hazard: Batteries used in today's conventional vehicles are lead-acid. Lead is a known carcinogen and extremely harmful to the environment.

According to an article from hybridcars.com,

More than 40,000 metric tons of lead are lost to landfills every year. According to the federal Toxic Release Inventory, another 70,000 metric tons are released in the lead mining and manufacturing process.

The batteries used in hybrid and electric vehicles are not without their environmental impacts as well. The materials that are used for the manufacture of the batteries themselves can contain, nickel, copper and other "rare" earth metals. The mining techniques normally used to get to these materials include strip mining, which leaves quite the literal hole in the landscape (image below). greentumble.com

And, as was pointed out in this article from *Discover Magazine*, Chinese miners, while mining for the lithium for magnets used in the batteries, often ignore environmental safeguards. The miners used acid after removing the top soil to extract the gold-flecked metals, and, in doing so, leaked acid into the groundwater destroying nearby agricultural land.

The production of the batteries themselves can have negative impacts on the environment as well. The production of the newer lithium-ion batteries account for 2 to 4 percent of a hybrid's lifetime emissions. The nickel-hydride batteries used in older hybrids release about 22 pounds of sulfur oxide emissions per vehicle which is 10 times more than a conventional powered vehicle.

As with anything, batteries will eventually need to be replaced. What happens to the batteries at this point?

Well, the folks over at the automotive site Emunds.com have a great article on the recycling of batteries.

Basically: The nickel-metal hydride batteries are "zero landfill" products, which means that once the nickel, copper, and rare metals are extracted, the rest of the battery is destroyed. There is literally nothing left to send to the landfill.

Author note: I do not understand the above comment. I refer to the statement, There is literally nothing left to send to the landfill" I assume when the valuable innards are taken out the case and other whatever containing the inner working parts there has to be something that is disposed of somewhere? Like the "slag" mentioned below. Just what is defined as slag I wonder and is it hazardous?

The newer lithium-ion batteries will typically still have 80% capacity after they stop working in an automobile. Power companies are working with automakers to potentially use them as energy storage for solar panels and wind turbines that produce energy during "off hours." The batteries are also being evaluated as backup power storage systems for retail and residential applications. Even if they can't be used in this aspect, there is a process to remove all the metals which leaves a kind of slag leftover. This slag can then be sold as an aggregate to concrete companies to make concrete stronger.

Author note: I wasn't born yesterday and have an inquisitive mind. I queried the above concrete statement, and I want you to know that it is not true! During my research, I discovered the following information online. Who are they trying to support the mighty buck? They are totally screwing around with the environment!

Yes, battery recycling slag, especially from lead-acid batteries, can be hazardous when used in concrete slabs. This is because the slag may contain heavy metals like lead and other toxic substances that can leach out and contaminate the surrounding environment.

Here's a more detailed explanation:

- **Hazardous Waste:**

  Battery recycling processes, particularly for lead-acid batteries, are regulated as hazardous waste activities. This is because the process involves draining electrolytes, crushing, smelting, and other methods that can release harmful substances into the environment.

- Leaching **of Toxic Elements:**

  The slag produced during recycling can contain heavy metals like lead, which can leach out into the surrounding environment if not responsibly managed. This leaching can contaminate soil and water, posing a health risk.

- Environmental Risks:

  Using battery slag in concrete can lead to environmental risks, including the contamination of water resources and the release of toxic elements into the air.

- Proper Management:

  To mitigate the hazards, battery slag should be responsibly managed, including storage and disposal, to prevent leaching and other forms of contamination.

- Recycling Alternatives:

  There are established recycling processes for battery waste. Professionally managed recycling facilities can extract valuable metals from batteries and minimize the environmental impact of waste generation.

What do you think?

The very outlet that a plug-in hybrid or electric vehicle plugs into can make or break its environmental impact. The manufacture process, materials used, and retrieval of those materials though strip mining impact the environment. It also depends on what state you live in. There are pros and cons to the purchase and use of hybrid cars, so ultimately, the question remains: how efficient are hybrid and electric vehicles at protecting the environment?

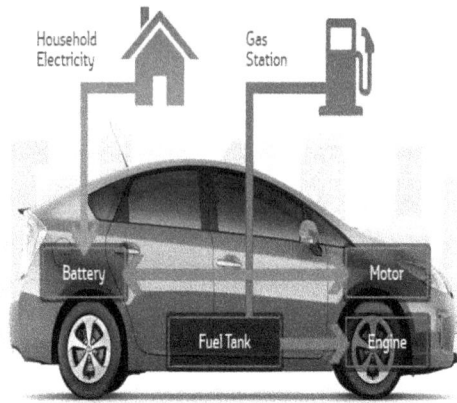

Hybrids: A Harmful "Green" Myth

While we have previously and extensively written about the myth that hybrid vehicles are in some way "green" or "sustainable" a new report published today has given us cause and further evidence to dispel this illusion.

How harmful are hybrids?

A collaborative study by Transport and Environment and Greenpeace has found that carbon dioxide emissions from plug-in hybrid vehicles are up to two-and-a-half times greater than official tests and manufacturer marketing suggests.

While official tests record plug-in hybrids as emitting an average of 44g of CO_2 per km driven on closed circuit conditions, today's report found that the same vehicles produced 120g of CO_2 per km when driven on real-world roads.

This data was gathered from over 20,000 plug-in hybrid drivers around Europe and demonstrated that the average plug-in hybrid car would produce around 28 tonnes of CO_2 in its lifetime – only offering a small reduction on the 39-41 tonnes of CO_2 produced by the average petrol or diesel car, or the 33 tonnes produced by non-plug-in hybrids.

Rebecca Newsom, Greenpeace's Head of Politics, stated the following when speaking to the BBC: "They may seem a much more environmentally friendly choice, but false claims of lower emissions are a ploy by car manufacturers to go on producing SUVs and petrol and diesel engines."

Separating marketing from reality

To better understand where the difference in these figures comes from, we need to look how emissions tests are managed, and who they are run by.

Typically, they will be conducted with the plug-in hybrid on a full charge and run at low speeds to ensure that the car operates using only its electric motor for the longest possible time before it switches to using its petrol/diesel engine once it reaches a certain distance or speed.

REALLY HOW GREEN IS GREEN?

This creates a more marketable result that car companies can use to promote their hybrid lines as "green" choice vehicles. However, it doesn't reflect how these vehicles will be used in the real world.

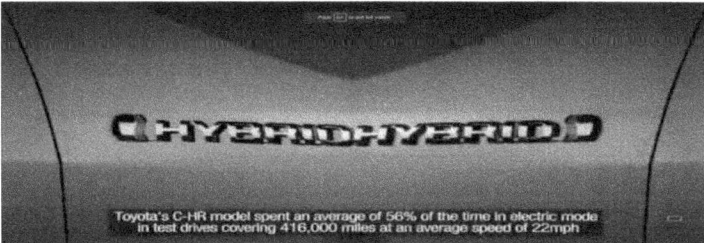

Screenshot of Toyota's C-HR advert

For example, Toyota's current TV advert for its C-HR carries a disclaimer in the small print, stating that their model "spent an average of 56% of its time in electric mode in test drives covering 416,000 miles at an average speed of 22mph".

When did you last travel at an average of 35khp over a distance of 416,000 miles?

This approach of misleading consumers by car manufacturers has been coupled with aggressive marketing to companies who operate large fleets, with the aim of those businesses being able to explicitly state that they are "going electric" or reducing their carbon footprint.

In reality it is commonplace to see many of these plug-in hybrids returned with their chargers still in their original packaging and completely unused – having been run solely on fossil fuels and offered no emissions reductions.

"A hybrid will be kicking out at least 40-70% of the emissions of a petrol or diesel car."

Wolves in sheep's clothing

While many hybrids are marketed as being "electric", this is incredibly misleading as they predominantly run on petrol or diesel for the majority of their use and therefore create harmful emissions that contribute to climate change.

Hybrid manufacturers are trying to tap into the incredibly high demand for fully electric vehicles by marketing them as something they are not.

With the impending ban on fossil fuel vehicles due to come into effect in Ireland in 2030, the development of hybrid vehicles allows manufacturers to effectively sell you two vehicles rather than one – a hybrid now (which will be banned in the near future) and then an EV when the ban on hybrids comes into effect.

In the meantime, the hybrid will be kicking out at least 40-70% of the emissions of a petrol or diesel car and will have created 15% more emissions in its manufacture than a battery electric vehicle would have, according to a study by the Low Carbon Vehicle Partnership in collaboration with Ricardo.

Furthermore, while battery electric vehicles can be charged at home and by renewable energy, hybrid vehicles still require fossil fuels. This means that they require oil to be extracted, refined and transported from around the world, creating untold emission levels in the refinement process and during transport to you – known as fuel miles.

And if that doesn't convince you, hybrids still require you to spend your hard-earned money on petrol or diesel – which is significantly more expensive than electricity.

Simply put, battery electric vehicles create no emissions themselves, while hybrids continue the same cycle that petrol and diesel vehicles do. Can't tell the difference? Just check to see if it has an exhaust – if it does, then it's a hybrid.

What's next?

We have long supported calls to ban the sale of hybrid vehicles in 2030, alongside the impending ban on the sale of new petrol and diesel vehicles.

Author's note:

Well, I guess you really need to evaluate your driving habits before jumping into the HYBRID pool. Mining and power generation can outweigh the benefits, giving a prospective buyer lots to think about concerning the value of the marginal benefits, and where power actually comes from to charge the car.

On another note, remember when I introduced you to companies that were experimenting with Hydrogen technology and had a method of producing fuel cells that could power car engines? Well, would you be surprised if I told you Hydrogen car engines do exist and that many companies are working on concepts for cars and trucks? Let's take a look at who and the technology.

The author reveals:

I have worked as an engineer with a company as its operations control VP in the early development of a hydrogen fuel cell, specifically the Venturi part, and its release on demand. We solved the issue and sold it to Ford, which, by the way, is working with the engine/fuel cell hydrogen applications.

While working with an educational association, The Altschuller Institute, as its international educational program manager, I invited a person, Dr. Paul McCreedy of Aero environment, who by the way was also the National Science Assoc. "Engineer of the 20th Century". By some quirk we became close friends, and I shared my experiences with him over time. I can only tell you some aircraft he designed has this special technology to do with hydrogen and that's as far as I can go.

Let's examine some programs/companies working with this zero-emission fuel.

Toyota Ammonia Engine 2024

Toyota and GAC's Groundbreaking Ammonia Engine Is Not an EV Killer

Published:

24 Oct 2023, 17:45 UTC

By:

Alex Oagana

Photo: GAC, edited by autoevolution

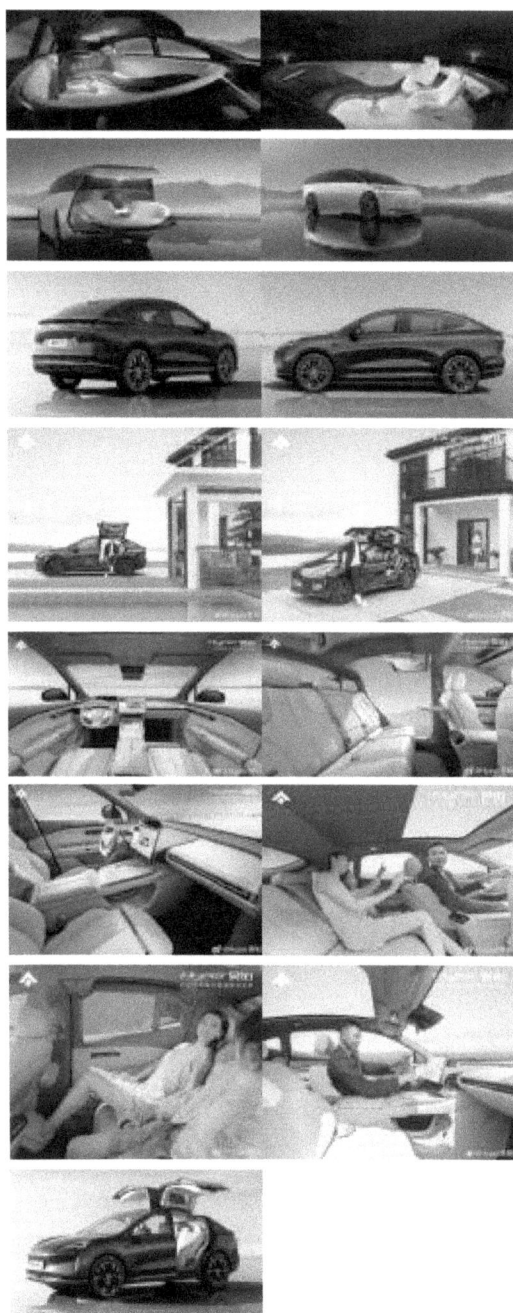

In a time when the global automotive industry is actively transitioning from internal combustion engines toward cleaner alternatives like EVs and, to a lesser extent, FCVs (Fuel Cell Vehicles), the state-owned Guangzhou Automobile Group (GAC) from China is taking a distinct approach.

At its annual technology showcase earlier this summer, the long-standing Chinese partner of Toyota unveiled an innovative internal combustion engine powered by ammonia, of all things. For those not in the know, which is probably most people, ammonia has been previously harnessed as a combustible fuel, particularly in the shipping and trucking sectors.

That said, GAC's pioneering use of this as a fuel in automotive applications managed to raise some eyebrows. At first glance, this signifies a unique attempt to explore alternative propulsion technologies beyond the conventional ICE systems without switching completely to BEVs (Battery Electric Vehicles), something that GAC is also known for. In other words, it's trying to do precisely what hydrogen and CO2-neutral gasoline proponents are trying, but is an ammonia engine really an alternative?

According to the press release, GAC's prototype ammonia engine is a 2.0-liter four-cylinder with 161 HP (163 PS) that doesn't seem to use any forced induction. Obviously, the high(ish) output per liter isn't the biggest story with this powerplant, but its 90 percent carbon emissions reduction rate compared to a fossil-fueled engine with similar capacity and power.

GAC hasn't mentioned anything about fuel economy, but we can assume it's nothing horrendously bad or particularly good. Either way, that is beside the point because we're here to talk about ammonia and how it could potentially become a death stroke for the proliferation of EVs, or can it? Ammonia is corrosive and engines will deteriorate.

What on Earth Is Ammonia?

Ammonia is a highly versatile chemical compound comprising nitrogen and hydrogen (NH3) and is renowned for its widespread applications. While primarily used as a fertilizer in agriculture, with over two-thirds of the world's annual production of ammonia being used for fertilizing crops, the substance's significance extends into a multitude of areas. Ammonia doesn't grow on trees, nor can it be pumped from the ground; it is a foundational chemical synthesis element and a key molecule in numerous processes across different industries. People much smarter than me say that ammonia is an intermediary in synthesizing sodium bicarbonate, nylon, synthetic fibers, plastics, polymers, and even explosives.

It also finds its place as an ingredient in products like paints, hair dyes, and household cleaners. Beyond this, ammonia serves as a refrigerant, a solvent, and a whitening agent in paper manufacturing. Its contributions extend into rubber manufacturing as a stabilizer and metallurgy as a valuable reducing agent.

Moreover, the substance acts as a reagent for controlling nitrogen oxides (NOx) in aqueous solutions, particularly in the exhaust systems of diesel engines. Due to its extensive utility, ammonia ranks among the most produced and utilized high-volume chemicals globally.

Now, with great power comes great responsibility, as Uncle Ben used to tell Spider-Man, and ammonia, in general, needs a lot of responsibility in its handling.

The Pros and Cons of Using Ammonia as Fuel

Ammonia holds the promise of being a zero-emission fuel due to its almost total lack of CO_2 emissions during combustion. Nevertheless, engines currently running on ammonia do produce exhaust gases, and these include unburned ammonia, nitrogen oxides (NOx), and nitrous oxide (N_2O), both of which is particularly good for your health. Quite the contrary.

Since the substance has no carbon atoms, CO_2 production levels from burning ammonia are negligible. That's all great, but then you need to start thinking about the energy and equipment required to transform it into fuel for a combustion engine, and that's when all those annoying variables begin to pop their ugly heads.

There are a few ways, some less viable than others, to make new gasoline ammonia, and none of them seem to offer the true benefits expected from an alternative fuel.

First of all, not unlike Porsche's synthetic e-fuel project, ammonia requires an insane amount of energy to be converted into fuel that can be used in cars. You can minimize the energy requirements by using part of the ammonia created to synthesize hydrogen since the substance is three parts hydrogen anyway. Then, you can use said hydrogen to create a fuel cell that produces electricity. But, then again, that would be like scratching a wooden leg, a suit of problems you don't need.

It's also similar to the way Porsche is creating synthetic gasoline. They capture carbon dioxide directly from the air and combine it with hydrogen to synthesize methanol. The resulting methanol is then used in a methanol-to-gasoline (MTG) process, resulting in a fuel that acts the same way as modern gasoline.

A more viable way to transform ammonia into a usable automotive fuel would be to create a so-called cocktail of various ingredients, including diesel, gasoline, or even hydrogen, to make a liquid fuel that can be used in internal combustion engines like any other fuel type.

Apparently, GAC's prototype engine, developed in partnership with Toyota, can be powered by liquidized ammonia ($NH3$) together with... something else. This used to be a technology used only on ships or farming equipment, so let's see why it's a bad idea for cars.

The Ugly Truth

Toyota has garnered attention over the years for its innovative alternative energy approaches and ambitious fueling objectives, especially compared to its somewhat slow jump on the EV bandwagon.

In other words, it's no surprise that GAC's prototype engine is said to have been developed with some of Toyota's expertise in searching for BEV alternatives. Unfortunately, there are so many red flags and negative variables surrounding liquid ammonia that even Porsche's e-fuels, which currently cost around $45 a gallon (or 10 euros per liter), make more sense financially.

Firstly, liquid ammonia is highly toxic and extremely corrosive, and on top of it, after combustion, it releases a lot of nitrogen into the atmosphere. These three reasons alone would make it non-viable as an e-fuel replacement or an EV-killer. Still, GAC and Toyota believe they have solved the nitrogen-emitting part by making the prototype engine with an exceedingly high compression ratio, among other things.

Secondly, the two remaining downsides are pretty much unsolvable, at least in a viable way. Can you even imagine the horror of dealing with the fuel system maintenance of an ammonia-running engine? What if that high compression isn't maintained because of some unforeseen problems with the rest of the car?

It would just pollute a thousand times more than any gas-guzzling vehicle. And by pollution, I mean more nitrogen in the atmosphere, which usually results in more ozone layers, respiratory illnesses, acid rain, and the works.

The corrosive nature of the substance also makes it tricky to manage before it even gets in your fuel tank (and after). In the minds of many, an ammonia engine is another cure that somehow seems worse than the purported disease it is trying to cure. We're probably better off with dealing with the infrastructure needed for charging all the millions of BEVs that are coming.

8 Vehicle Manufacturers Working on Hydrogen Fuel Cell Cars

Hydrogen's Promising Impact on Local Communities / July 7, 2023

We've come a long way since the early days of the automobile. From 18th-century steam-powered behemoths to gas-guzzling modern supercars, vehicles and their fuel sources have progressed throughout history. Recently, another shift has begun. Now, vehicle manufacturers are using hydrogen fuel cells to propel personal transportation forward.

Admittedly, the idea of a hydrogen-powered car is nothing new. More than 200 years ago, the first internal combustion engine ran on hydrogen and oxygen. But these new models are different. Hydrogen fuel cell electric vehicles (FCEVs) are ultra-efficient, converting pure hydrogen gas into electricity without producing any harmful tailpipe emissions.

Did you know about 15,000 hydrogen-powered vehicles are on U.S. roads right now, all in California? This carbon emission-free technology has caught the attention of some of the auto industry's biggest players. In this article, we'll look at some hydrogen car companies betting on hydrogen to drive sustainable mobility forward.

Toyota

Believe it or not, Toyota has been working on fuel cell vehicles since 1992. Toyota has been steadfast in developing fuel-cell vehicles—long before many of its counterparts even considered the idea. The Toyota Mirai was among the first mass-produced hydrogen fuel cars to hit the road in 2014.

Its introduction was a major milestone in automotive and sustainability history. As one of the world's first mass-produced fuel cell electric vehicles (FCEVs), the Mirai showcased hydrogen cell vehicles' potential. With a 312-mile driving range, the Mirai demonstrated that zero-emission driving was not a pipe dream but a reality.

In 2022, Toyota sold 2,094 Mirai, a 20% YOY increase over 2021. This sales growth proves a growing appetite for cleaner vehicles, especially as hydrogen infrastructure progresses.

And Toyota hasn't stopped there. According to CNBC: "Alongside the Mirai, Toyota has had a hand in developing larger hydrogen fuel cell vehicles. These include a bus called the Sora and prototypes of heavy-duty trucks. Alongside fuel cells, Toyota is looking at using hydrogen in internal combustion engines.

Toyota Hilux

In December 2022, Toyota announced its plans to develop a hydrogen fuel cell version of its best-selling Hilux pickup. Only a little is currently known about the new Hilux FCEV. However, with funding from the UK government and support from various industry leaders, the truck is destined for success. The hydrogen-powered Hilux will be ready for small-scale production if all goes well with prototyping.

Additionally, Toyota announced plans to produce hydrogen fuel-cell modules to replace heavy-duty diesel engines in Class 8 semi-trucks starting this year.

Toyota Yaris H2

Toyota introduced the GR Yaris H2, an advanced hydrogen-powered concept car, in December 2021. Unlike the Mirai, which converts hydrogen fuel into electricity, the GR Yaris H2 uses hydrogen as a combustible fuel, offering a traditional combustion engine experience.

It features a modified turbocharged three-cylinder engine from the standard GR Yaris and is expected to generate around 268 horsepower. While official figures are not yet available, industry experts believe it could match or even surpass the performance of the acclaimed GR Yaris. Although still in early development, the prospect of a hydrogen combustion vehicle from Toyota is exciting.

Toyota's Commercial FCEVs

Along with the Mirai and the new Hilux, Toyota has also been working on hydrogen buses and heavy-duty trucks. If larger transport vehicles transition to hydrogen alongside commuter cars, we might reach our zero-emissions goals even faster.

BMW

BMW has been in the hydrogen vehicle game for some time, with past projects including the i Hydrogen NEXT concept car and the 1 Series Fuel-cell hybrid electric. But in 2021, the German auto manufacturer announced its next focus: An all-hydrogen version of the BMW X5.

The BMW iX5 Hydrogen pilot fleet is now on the roads for trial and demonstration. The iX5 Hydrogen represents a significant milestone in the company's pursuit of eco-friendly transportation, integrating advanced technologies to deliver efficient and emissions-free driving. By leveraging the power of hydrogen, BMW aims to address the challenges associated with electric vehicle (EV) infrastructure and range limitations.

Specifications include:

- **401 horsepower**
- **Maximum speed:** 118 mph
- **Estimated driving range:** 504 miles.

We're excited to learn more and see the iX5 Hydrogen in action when BMW finally deems it road ready.

Hyundai

Hyundai Motor Company threw its hat in the hydrogen ring with the first fuel cell-powered SUV. In 2018, the South Korean vehicle manufacturer unveiled the NEXO Fuel Cell. Now in its fifth year of production, the Hyundai NEXO has helped eliminate more than 14 million miles of vehicle emissions.

NEXO's specifications include:

- **Driving range:** 380 miles
- **Combined MPGe:** 61
- **Maximum speed:** 111 mph
- **Hydrogen tank capacity:** 41.4 gal
- **Lithium-ion battery capacity:** 1.56 kWh

Hyundai plans to reveal significant enhancements for the 2024 Nexo. These changes are expected to include improved efficiency, thanks to implementing a third-generation fuel cell stack and an impressive range of nearly 500 miles.

Additionally, the 2024 Hyundai Nexo will likely showcase advanced safety and driver assistance features, enhanced interior design, and technological advancements.

Hyundai N Vision 74

The Hyundai N Vision 74 is a hydrogen hybrid concept that pays homage to the 1974 Pony Coupe Concept. The hybrid powertrain combines a lithium-ion battery and hydrogen fuel cell, delivering impressive acceleration. It's a futuristic marvel that will turn heads and attract buyers.

- **Driving range:** 372 miles
- 670 horsepower (combined)
- 800-Volt fast charging in just five minutes
- **Maximum speed:** 155 mph

Hyundai's Commercial FCEV

While hydrogen fuel cell technology can transform personal transportation, it also has commercial potential. In 2020, Hyundai released the XCIENT Fuel Cell—an achievement it calls "the world's first mass-produced heavy-duty fuel cell truck line."

The XCIENT is a zero-emissions cargo vehicle with more than 400 km of driving range. Already on the road in Switzerland, it'll likely transform the overland shipping industry as it reaches more locations.

Honda

Honda is no stranger to hydrogen fuel cell vehicles. Between 2017 and 2021, the Japanese motor company sold a hydrogen-powered version of its Clarity model before discontinuing it due to lackluster sales.

However, in November of 2022, Honda announced plans for its electrification push, which included US-oriented hydrogen fuel cell cars. The automaker plans to begin producing fuel cell vehicles in the states starting in 2024 with a plug-in hybrid CR-V.

At the time of writing, details on the zero-emissions CR-V—set to be released in 2024—are relatively slim. However, here's what we do know:

- **Plug-in charging meets hydrogen fuel cell technology** – Honda has announced that the CR-V will be the first FCEV in the North American market to incorporate a plug-in capability for electric vehicle (EV) driving. This will enable motorists to fuel up at home and on the go—without worrying about the current lack of hydrogen refueling stations.

- **A familiar look** – Aside from different badges and trim options, the new CR-V is expected to resemble the standard version.

- **Domestic production** – The hydrogen-powered CR-V will be produced at Honda's Performance Manufacturing Center in Marysville, OH.

Next year will tell how Honda plays a hand in advancing hydrogen energy technology—and how it scales alongside its competitors.

In Development

Here are a few other FCEVs to keep on your radar, currently in development.

Riversimple Rasa

The Rasa is a concept car that can travel 300 miles with just 1.5 kg of hydrogen, surpassing the average hydrogen car range. It accelerates from 0 to 60 mph in 9.7 seconds and features a unique ultracapacitor system for improved energy storage.

After debuting in 2016, it will finally enter production in 2023 to manufacture 5,000 units per year. Even King Charles III of the UK test-drove it! The Rasa presents an exciting and practical solution for hydrogen-powered personal transportation.

Land Rover Defender FCEV

Jaguar Land Rover aims for zero tailpipe emissions by 2036. This initiative focuses on creating a hydrogen fuel cell vehicle based on the Land Rover Defender SUV.

While specific range and power details are undisclosed, the vehicle features electric drive units, a fuel cell stack, a large battery for energy recuperation, and a high-pressure hydrogen tank. The Land Rover Defender FCEV's public availability timeline remains to be determined, but its potential is promising.

NamX HUV

The NamX HUV, a hydrogen concept car from Pininfarina, aims to revolutionize driving. It features a permanent H2 tank and a removable setup of six capsules, which could greatly boost the adoption of hydrogen vehicles. NamX plans to offer home capsule delivery and establish new hydrogen capsule refueling stations to address refueling concerns.

Development is scheduled to start in 2023, with a release expected by the end of 2025, priced between $80,000 and $100,000. The NamX HUV possesses unique features that could disrupt the market and usher in a new era of hydrogen vehicles.

Hyperion XP-1

Hyperion Motors, an American manufacturer based in Columbus, Ohio, created a remarkable hydrogen-powered car called the Hyperion XP-1. This hyper car made headlines upon its unveiling in August 2020.

It boasts advanced technology, featuring an innovative hydrogen powerhouse engine module with storage technology derived from NASA. The Hyperion XP-1 offers an astonishing range of over 1,000 miles and produces a staggering total power output of over 2,000 horsepower.

Closing Note

As we wrap up this first volume, I find myself reflecting on what the New Green Deal really means in practice, beyond slogans and policy debates. We've explored hydrogen-powered vehicles not just as an engineering curiosity, but as one of the many real-world steps that could help turn big promises into everyday reality. It's encouraging to see innovation at work, yet it also reminds me that vehicles are just one piece of a far larger puzzle.

The New Green Deal is, at heart, about how we live, produce, and power our world differently. That journey doesn't end at the showroom floor or the refueling station — it extends into the systems that deliver energy, the trade-offs we weigh, and the choices we make about cost, reliability, and fairness.

In the next volume, we'll look beyond new technologies alone, and ask: what does it really take to build the infrastructure, energy networks, and policy frameworks that could make the New Green Deal work.

www.ingramcontent.com/pod-product-compliance
Lightning Source LLC
Chambersburg PA
CBHW052110030426
42335CB00025B/2917